LEARN MATH FAST SYSTEM VOLUME V

By J.K. Mergens

Learn Math Fast System Volume V
Copyright © 2014, 2018, 2019, 2021
ISBN: 978-1530513277
www.LearnMathFastBooks.com

United States coin image from United States Mint

TABLE OF CONTENTS

Introduction 5

PRE-ALGEBRA Review Test 1 7

Chapter 1: Algebra Lingo 9
- Lesson 1: Integers and Natural Numbers 9
- Lesson 2: Rational and Irrational Numbers 12
- Lesson 3: Variables and Coefficients 15
- Lesson 4: Terms, Like Terms and Expressions 19
- Lesson 5: Factors and Absolute Value 23
- Chapter 1 Review Test 27

Chapter 2 Exponents[1] and Terms with Exponents[2] 28
- Lesson 6: Combining Like Terms 28
- Lesson 7: Simplifying Expressions 32
- Lesson 8: Multiplying with Exponents 35
- Lesson 9: Dividing with Exponents 38
- Lesson 10: Dividing terms with exponents 46
- Lesson 11: Dividing Expressions by a single term 51
- Lesson 12: You Can Only Cancel Factors – Not Terms 60
- Lesson 13: Squared and Cubed 68
- Lesson 14: Square roots 73
- Lesson 15: Cube Roots 75
- Lesson 16: Square Root of a Fraction 80
- Chapter 2 Review Test 82

Chapter 3 Laws of Algebra 84
- Lesson 17: Laws of Addition 84
- Lesson 18: Laws of Multiplication 88
- Chapter 3 Review Test 91

Chapter 4 Distributive Law of Multiplication 93
- Lesson 19: Multiplying a Term by an Expression 93
- Lesson 20: Distributing a Negative Number 96
- Lesson 21: Multiplying a Term by a Trinomial Expression 100
- Chapter 4 Review Test 106

Chapter 5 Factoring and Prime Factorization 107
- Lesson 22: Factoring 107

- Lesson 23: Prime numbers 110
- Lesson 24: Prime Factorization 112
- Lesson 25: Factoring Out a Common Factor 114
- Chapter 5 Review Test 120

Chapter 6 Polynomials 123
- Lesson 26: Monomials, Binomials, and Trinomials 123
- Lesson 27: Multiplying Binominals 128
- Lesson 28: Multiplying with Trinomials 137
- Lesson 29: Factoring A Quadratic 142
- Chapter 6 Review Test 150

Chapter 7 Quadratic EQUATIONS 152
- Lesson 30: Recognizing a quadratic equation 152
- Lesson 31: Solving for x in a Quadratic Equation 155
- Lesson 32: A Perfect Square 160
- Lesson 33: Completing the Square 167
- Lesson 34: The Quadratic Formula 182
- Lesson 35: Intro to Functions 187
- Lesson 36: Final Review 193
- Final Test 195

ANSWERS 199

INTRODUCTION

Algebra can be quite fun. You get to solve these big, complicated problems that most people can't even understand, but you can solve them easily. Look at this problem:

$$3x^2 + 3y + 1 + 5x^2 - 8x^2 + 2y - 5y + 1 =$$

Anyone who knows algebra can quickly solve this problem in their head and tell you the answer is 2. It's a lot like reading. A five-year-old would be confused by this word:

Photograph

He may see Pot-o-grap, but you see Fo-toe-graf. You instantly know this word means "a picture" because you know all the *rules* to reading. Algebra is the same way, once you learn the *rules*, it is easy to read any algebra problem.

Mastering these rules now will give you the ability to advance to higher math because the rules never change! What you learn in this book will still apply in Algebra II, Calculus and Physics. It all starts with learning the basic algebra skills explained here. At the end of the book, you will be amazed at the complex math you will be able to understand.

Before we start Algebra, let's make sure you have learned all the Pre-Algebra skills necessary to continue. If you can correctly answer the problems on the Pre-Algebra Review Test, you are ready to continue. If you have trouble with this test, you need to read Volume II and/or III of the *Learn Math Fast System*.

PRE-ALGEBRA REVIEW TEST 1

Solve the following.

1. $-7 + -7 =$

2. $5 + -8 =$

3. $(7) - (-3) =$

4. $6 + -10 =$

5. $-\frac{3}{8} + \frac{7}{24} =$

6. $\frac{2}{7} - -\frac{5}{21} =$

7. $-10 - 6 =$

8. $7 \times -6 =$

9. $-5 \times -5 =$

10. $15 \div -3 =$

11. $-24 \div -8 =$

12. $-48 \div 6 =$

13. $-\frac{1}{2} \times 2 =$

14. $-.07 \times -.7 =$

15. $-9.3 \cdot 4.1 =$

16. $\frac{3}{4}\left(-\frac{5}{8}\right) =$

17. $\frac{\frac{1}{3}}{\frac{1}{2}} =$ (Divide the fractions)

18. $-\frac{4}{8} \div \frac{3}{4} =$

Solve for x.

19. $10 + x = 13$

20. $x - 10 = 90$

21. $5x = 30$

22. $\frac{55}{x} = 5$

23. $\sqrt{64} = x$

24. $3^2 \div 3 + 6 = x$

Solve for x in terms of y.

25. $x - 1 = y$

26. $x + 2 = y$

27. $5 + x = y$

28. Use the slope formula to find the slope of a line with coordinates $(2, 3)$ and $(3, 5)$.

$$m = \frac{y_2 - y_1}{x_2 - x_1}$$

The answers are in the back of the book under Pre-Algebra Test 1. Check your answers, to see if you are ready to continue.

CHAPTER 1: ALGEBRA LINGO

LESSON 1: INTEGERS AND NATURAL NUMBERS

In the world of math, there are a lot of new words to learn. If you don't know the meaning of these words, it is going to be nearly impossible to understand algebra instructions. Let's take some time right now and go over some of the terminology. But don't worry, there are only 12 new words/phrases to learn and some of them you may already know, so let's get started.

Do you remember learning about odd and even numbers?

When you separate numbers like this into their own special groups, each group is called a *set of numbers* or a *number set*. In algebra, there are a few number sets you should learn about. I'll start with the biggest set of all; the set that contains every single number in the world. Every number under the sun is included in this set of numbers. It is the granddaddy of all number sets and they are called *Real Numbers*.

All numbers are considered to be *real numbers*, (except imaginary numbers, but you don't need to worry about those until college level math). It doesn't matter if it is odd, even, positive or negative, a decimal number, or a fraction – all numbers are real numbers.

Word #1 Real Numbers. Write that down.

When we count, we don't use every single real number. We count using only the *natural numbers*. Natural numbers are the numbers you use to teach a child how to count, you know, 1, 2, 3, 4, and so on. You wouldn't teach a child to start with 0 and you wouldn't teach negative numbers or use fractions either. You would just stick to the *natural numbers* 1, 2, 3, 4, 5, 6, 7...it's only natural!

Word #2 Natural Numbers, write that down too.

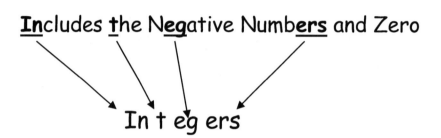

Once you include negative numbers and zero to all the natural numbers, you have a new set of numbers called *integers*. It's easy to remember which numbers are called *natural numbers* because that's how we *naturally* count, but the word "integer" isn't as obvious. To help you remember which set of numbers are called integers, use this trick.

Includes the Negative Numbers and Zero

In t eg ers

Keep in mind that fractions are not included in the sets of integers or natural numbers. Only whole numbers are included in these two sets.

Word #3: Integers. Write that down.

It is possible for a number to be both an integer and a natural number. For example, the number 3 is both an integer and a natural number. The number -3 is an integer, but it is not a natural number. A fraction, such as $\frac{3}{4}$, is neither an integer nor a natural number, but all of them are *real numbers*.

Name_____ Date _____

Worksheet 1

Name each set of real numbers as all natural numbers, all integers, or neither.

1. 5, 7, 18, 29
2. -2, 1, 3, 9
3. 0, 1, 2, 3

4. .5, 1, 1.5, 2, 2.5
5. -20, -21, -22
6. 1015, 1025, 1046

7. 1, ½, 3, 5
8. -1, 0, 1
9. 253.5, 600.5, 3

10. 7, 12, -3, 0, 5
11. 14, 12, 110, 6
12. ½, .335, $\sqrt{37}$, 0

13. Fill in the blanks to spell out the helpful integer phrase.

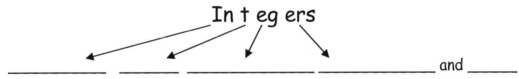

_____ _____ _____ _____ and _____

14. When you teach a child how to count, which set of numbers do you use?

15. Which set of numbers includes every number in the world?

* Review

16. $-3 \times -9 =$
17. $\frac{1}{4} \div \frac{3}{5} =$
18. $10 + x = -19$
19. $\frac{49}{7} = x$

20. $3a = 99$
21. $\frac{1}{2} \cdot \frac{1}{2} =$
22. $\sqrt{121} =$
23. $\frac{1}{2}x = 32$

24. $12^2 =$
25. $3\frac{4}{7} \times 4\frac{3}{5} =$
26. $4^2 + (3 - 1) + 5(8 + 2) =$

LESSON 2: RATIONAL AND IRRATIONAL NUMBERS

You've just learned the first three new words/phrases: real numbers, natural numbers, and integers. There are just two more sets of numbers to learn about, but don't worry, they are easy because their names make sense... a little bit.

We already said that ALL numbers are real numbers. Some of them are integers and some of them are natural numbers. Next, you will learn that every real number is either a *rational number* or an *irrational number*. Let me explain.

One definition of the word *rational* is "sensible" or makes sense. A rational person makes more sense than an irrational person, just like a *rational number* makes more sense than an *irrational number*.

Most numbers are rational, but every so often you will run into one of those numbers that have never ending decimal places. For example, you probably remember learning that the number for pi is 3.14, but that is a rounded number. Pi is actually a big, long, irrational number that never ends. Here are the first 31 decimal places.
$$\Pi = 3.1415926535897932384626433832795...$$

Since this number never ends or repeats, it is called an *irrational number*. An irrational number is impossible to write as a fraction, unless you round the number. But then it is not exact and that's the problem with irrational numbers. Here are two more common examples of irrational numbers.

$$\sqrt{2} \qquad \sqrt{7}$$

It is easy to solve a problem like $\sqrt{4}$ because the answer is a rational number, 2. But $\sqrt{2}$ and $\sqrt{7}$ have never ending decimal places.

$$\sqrt{2} = 1.414213562373095...$$
$$\sqrt{7} = 2.645751311064591...$$

The square root of 7 is 2.645751311064591...this number goes to the 100 trillionth decimal place and it's still not done. Since this number won't end, I cannot write it as a fraction. This number makes no sense...it is completely irrational.

Any number that cannot be written as a fraction is called irrational. Look at these next two numbers. It *is* possible to write these numbers as a fraction.

$$\sqrt{16} \qquad 53.007$$

The square root of 16 is 4, as a fraction that is written as $\frac{4}{1}$. The other rational number is 53.007, as a fraction that number is written as $\frac{53,007}{1000}$.

The difference between a rational number and an irrational number is that a rational number can be written as a fraction. The root word for rational is "ratio." A fraction is a ratio; that's the real reason why a number that can be written as a fraction is called a "ratio-nal" number.

I should also tell you that technically each number of the fraction must be an integer and the denominator can't be zero; that's the technical stuff. Basically, rational numbers are exact and irrational numbers have no end; that's all you really need to know.

Word #4 Rational numbers.
Word #5 Irrational numbers.

$\frac{3}{8}$

A Ratio

Name_____ Date _____

Worksheet 2

Name each real number as an integer or a natural number and then say whether it is rational or irrational. The first one is done for you.

*   17   A rational, natural number and an integer.
1.   -17
2.   .34
3.   .33
4. $\sqrt{4}$
5. $\sqrt{2}$
6.   -7/8
7.   0
8.   π
9.   100
10.   1/3

Review pre-algebra by solving the following.

11.   3a = 27   12. $\frac{1}{2}a = 5$ 13.   -12 x -12 =

14. $-8^2 =$ 15. $\sqrt{36} =$ 16.   x - 5 = - 10

17.   8(6 + 3) =   18.   (3 + 3) - 2^2 + (4 x 6)($\sqrt{4}$) =

Fill in the blanks with the correct order of operations (PEMDAS).

19. _____, Exponents, _____, _____, Add, _____

20. $-\frac{5}{9} \div \frac{5}{6} =$ 21. $\frac{7}{12} \times -\frac{8}{11} =$ 22. $2\frac{7}{9} + 4\frac{4}{27} =$

LESSON 3: VARIABLES AND COEFFICIENTS

You should be making a list and adding each new word you learn along with a short explanation, to help you remember the meaning. So far, you have learned 5 of the 12 new words: real numbers, natural numbers, integers, rational and irrational numbers.

The next word you probably know already, but I want to make sure. Look at the problem below; there is a number and a letter.

$$5a =$$

The letter is called the *variable* because it "varies." Any letter in an algebraic problem is a variable. Look at this next problem.

$$6m + 7m =$$

What is the variable in this problem? The letter m is the variable. The letter m could represent any number; however, every m in this problem must be the same number. It is possible to have more than one variable in a problem. Look at this next problem.

$$10x - 3a =$$

What are the variables in this problem? The letters "x" and "a" are the variables. Each variable represents a different number; that's why there are two variables in this problem.

I'll give you an example of how someone might use the formula above, $10x - 3a$. Let's pretend you get paid 10 dollars an hour to wash dishes, but for every dish you break, you lose 3 dollars. So, in the formula above, "x" will stand for the number of hours you worked and "a" will stand for the number of dishes you break.

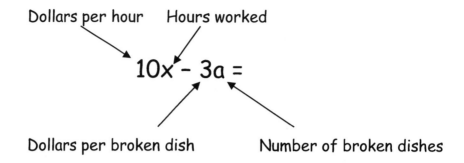

Fill in "x and a" with the appropriate numbers, to find out how much you get paid. Let's say you worked 8 hours and you broke 2 glasses. Use the formula to figure out your pay.

$x = 8 \quad a = 2$

10(8) – 3(2) means
(10 x 8) – (3 x 2) or
80 – 6 =

$10x - 3a =$
$10(8) - 3(2) =$
$80 - 6 = 74$ You get paid $74.00

That's how variables work! Each letter represents some unknown number. With this formula, you could calculate your pay each day. Just fill in "x" and "a" with the number of hours you worked and the number of dishes you broke.

Now that you know what the letters are in this problem, let's take a look at the numbers. Do you know what the numbers are called?

$10x - 3a =$

The numbers next to the variables are called the *coefficients*. The number 10 is the coefficient of x and -3 is the coefficient of a. *Coefficient* is a really big word to describe something as simple as the number in front of a variable. Name the coefficient of x in this next problem.

$3x + 4m =$

The coefficient of x is 3. Pretty simple, isn't it? Take a look at this next one.

$$a + 4b =$$

What is the coefficient of a? Even though you can't see a coefficient in front of the letter a, there is still an invisible 1. There are a lot of invisible 1's in algebra because any number times 1 is still that number. Just like an elephant is the same thing as 1 times an elephant.

Do you see how a variable all by itself is the same thing as a variable with a coefficient of 1? Read the examples below.

$1a = a$ $1 \times 2 = 2$ $1m = m$ $1 \text{ car} = \text{car}$ $1x^2 = x^2$

Write down the two new words you have learned in this lesson. Add them to your list along with a clue to help you remember each word.

Word #6 Variable.
Word #7 Coefficient.

Name_____ Date _____

Worksheet 3

Name the variable in each problem.

1. $5a - 7 =$
2. $a + 2 =$
3. $5b =$
4. $10a + 4a =$
5. c

Name the coefficient of y in each problem.

6. $3y + 7 =$
7. $4a - 2y =$
8. $\frac{1}{2} y =$
9. $y + 1 =$
10. $abc + 6y =$

Look at this problem then answer the following questions about it.

$$-8b + 3 =$$

11. Is the coefficient of b an integer?
12. Is the coefficient of b a natural number?
13. Is the coefficient of b an irrational number?
14. Is the number without a variable a rational number?
15. What is the variable?

Review.

16. $-\frac{4}{7} \times \frac{4}{7} =$ 17. $-\frac{4}{7} \div \frac{4}{7} =$ 18. $-\frac{4}{7} - \frac{4}{7} =$

LESSON 4: TERMS, LIKE TERMS AND EXPRESSIONS

Be sure to write down all 7 words that you've learned so far. Give each one a short explanation, to help you remember each word or phrase.

Next, we will learn what the word *term* means in algebra. This word is one of the most important words on the list because we use it a lot! Any number is a term. I'll start with the number eight. The *value* of this term is 8. The *value* is how much it is worth.

$$8 \leftarrow \text{This is a term}$$

I will add a variable to this number, it is still a term. The value is now 8 times some number that we will call "y."

$$8y \leftarrow \text{Term}$$

I will build up this term a little more.

$$8y^2z^2 \leftarrow \text{Still a Term}$$

I can multiply this term by 2 and it will still be a term.

$$2yz(8y^2z^2) \leftarrow \text{Still a Term}$$

But once I **add or subtract** *another term from this term* with a plus or minus sign, it becomes an *expression*.

$$2yz(8y^2z^2) + 10a \leftarrow \text{An Expression}$$

An *expression* is created once you hook terms together with plus or minus signs. You must know the difference between a term and an expression to excel in algebra.

So, do you understand what a term is? If not, you should go back and read this lesson again because the next phrase you are going to learn is *Like Terms*. So, if you don't know what a *term* is yet, this next phrase may confuse you. But if you understand that a *term* is a separate value not added or subtracted to any other term, then read on.

Look at the two terms below; they sort of look alike. They have the same variables and exponents, but the coefficients are different.

$$3xy^2 \qquad 7xy^2$$

It looks like they have the same last name, but they have different first names. These are called *like terms*. They can be combined because they are alike. I say "combined" because they might be added together, or they might be subtracted. I don't know which it will be, so I just say it's OK to *combine* them.

In order for two terms to be *like terms*, the coefficients do not need to be the same. Only the variables and exponents need to be the same. Here is another example of two like terms that can be combined.

$$9ab^3c - ab^3c =$$

This subtractions problem is asking us to subtract one ab^3c from nine ab^3c's. They have the same last name, so they are like terms. You will learn how to add and subtract terms later in this book. Right now, you are just learning how to tell the difference between like terms and terms that are not alike. Can you tell if these two terms can be added together?

$$5m^3n^3 + 4m^3n^3 =$$

Yes, they can.

Can these two?

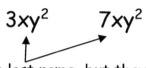
The coefficient of "a" is an invisible 1.

$$200a + a =$$

Yes, those terms can be combined too; they are like terms. I will make one slight change to each one of those same examples, below. Now they are unlike terms; they cannot be combined. See if you can find the difference that makes them not like terms.

$$3x + 7xy =$$
$$9ab^3c - ab^2c =$$
$$5m^3n + 4m^3n^3 =$$
$$200a + a^2 =$$

None of these problems can be solved any further.

Get out your list of Algebra Lingo and add these 3 words to your list. Give yourself an example too; one that will help you remember the meaning of each word.

Word #8 Term
Word #9 Expression
Word #10 Like Terms

Complete the next worksheet.

Name_____ Date _____

Worksheet 4

Name each of the following as either an expression or a term.

1. $4y$
2. abc
3. $12ab^2$
4. $3a + 4b$
5. $3^2 - (5a + 2c)$
6. $6y^2 + 8$
7. $1/2ab(3a)$
8. $4 + 3xyz$
9. $9a$
10. $d\pi$

Fill in the blanks with a word from your list that matches the description.

11._____ A number set that includes all negative and positive whole numbers including zero.

12._____ The set of numbers that we use to count.

13._____ The set of numbers that contain every number.

14._____ The type of numbers that can be written as a fraction.

15._____ The type of numbers that cannot be written as a fraction.

16._____ Terms hooked together with a plus or minus sign.

17. Are there any like terms in the problem below? If so, what are they?
$$7xy + abc + 4xy + 5abc + 7xy^2 =$$

LESSON 5: FACTORS AND ABSOLUTE VALUE

The next word is the MOST important word on the list. I'm sure you've heard of a *factor* before, but I'm going to make sure you FULLY understand the meaning of a *factor* because you cannot excel in algebra, unless you completely understand factors. Look at this multiplication problem.

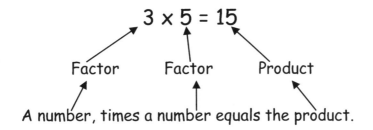

Factors are the numbers being multiplied. Can you name the factors of this next term?

$$7ab$$

Think about what is being multiplied…7 x a x b, so those are the factors of 7ab. It's just that simple. Even if the numbers being multiplied look like this next problem, they are still called factors.

$$(5x + 3)(2x + 3) = 10x^2 + 15x + 6x + 9$$

Factor   Factor   Product

When two sets of parentheses are next to each other, like this next problem, it means to multiply the two sets.

$$(5x + 3)(2x + 3)$$

Since this is a multiplication problem, the factors are (5x + 3) and (2x + 3). The factors are the "numbers" being multiplied. In this case (5x + 3) is being multiplied by (2x + 3), so those are the factors. Look at the next problem.

3 + 2

There are no factors in this problem because nothing is being multiplied. Factors are the two "things" being multiplied together. I say "things" because it could be two numbers, two variables, two sets of parentheses or a coefficient and a variable. If they are being multiplied together, then they are the factors of the answer.

Later in this book you will learn about *factoring*. Factoring is when you are given the product and then you have to figure out what factors were used to get that product. For now, you just need to know what makes a number a factor.

One more phrase to learn, then we will get to the fun stuff. It's one of the easiest phrases to learn in Algebra, everyone gets it. Let me explain. Some numbers are negative and some are positive. When you drop the sign, you are left with the *Absolute Value* of that number. For example, the absolute value of -7 is 7. The absolute value of +23 is 23. The absolute value of -95 is 95. Pretty simple huh?

There is a symbol for absolute value; it looks like this:

$$|-7| = 7$$

That sentence reads, "The absolute value of negative seven, equals seven." Can you tell me the answer to this problem?

$$|-8| + |10| =$$

That's easy, it's 18. These bars mean "ignore the sign," so the math is just 8 + 10. This is useful when you are measuring certain distances. For example, if a robot travels six feet forward and then six feet backwards, it might look like he didn't travel at all because 6 feet + negative six feet equals zero. If we use the absolute value, the robot will have traveled 12 feet. We won't deal with absolute value much in this book, but you should be aware of the symbol and what it means.

Add the last two words/phrases to your list. Make sure you understand all 12 new words on your list. Complete the next worksheet for practice.

Word #11 Factor
Word #12 Absolute value

Name_____ Date _____

Worksheet 5

Separate the following terms into their factors. The first one is done for you.

Term/Expression	Factors
1. $7y^3$	7 y y y
2. $3xy^2$	
3. $5x$	
4. a^4	
5. $11xy^2z$	
6. 3^3	
7. 7×2	
8. $8 + 5$	
9. $3a$	
10. $21a^3b^2$	
11. $(3 + 2)(4 + b)$	
12. 10^2	
13. $3(a + b)$	
14. $a(2 + a)$	
15. ab	

Solve the following problems.

16. $|8| + |-9| =$ 17. $|-3| + |5| =$ 18. $|7| - |-3| =$

19. $|-12| - |-5| =$ 20. $|-25| + |5| =$ 21. $|-9| + |3| + |-6| =$

22. $|-5| \times |9| =$ 23. $|7| \times |-8| =$ 24. $|-4| \times |4| =$

Name_____ Date _____

CHAPTER 1 REVIEW TEST

Name each number as either a natural number, an integer, both or neither.

1. 5   2. -5   3. $\frac{1}{5}$ 4. $-\frac{1}{5}$ 5. 0

Name each number as rational or irrational.

6. 3/8   7. $\sqrt{49}$ 8. $\sqrt{2}$ 9. π 10. 2

Name the variable and the coefficient of each term.

11. 2a   12. b   13. $10x^2$ 14. $\frac{5}{8}$ m   15. πr^2

Name each of the following as either a term or an expression.

16. $3xy^2$ 17. 8a(3y)   18. $\sqrt{9} + \sqrt{7}$ 19. 6a + 12y   20. a

Look at the expression below. Circle all like terms.

21. $3a^2b^2 + 7ab - a^2b^2 + 4ab^2 - 7a - 6a^2b^2 + 5a^2b + 9a^2b^2c =$

List all the factors of the following terms.

22. 5ab   23. $7y^2$ 24. 11abc   25. (5a + 3)(2a + 4)   26. $(a + b)^2$

Solve the following.

27. |-9| =   28. |8| - |-3| =   29. |27| * |-2| =

CHAPTER 2 EXPONENTS[1] AND TERMS WITH EXPONENTS[2]

LESSON 6: COMBINING LIKE TERMS

The first algebra skill you must master is working with exponents. You already know that 3^2 means 3 x 3 or 9. Now you will learn how to add, subtract, multiply and divide numbers and variables with exponents too. I have good news though – it's really simple to learn all these skills in just a day or two.

We will use 4^3 as our first example. In this term, the number 4 is called the base, and the number 3 is the exponent.

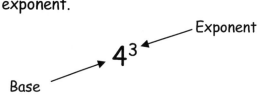

The exponent tells us how many times to use the base number as a factor in a multiplication problem. The term 4^3 means use the number 4, three times in a multiplication problem. $4^3 = 4 \cdot 4 \cdot 4$ or 64. The term x^3 means $x \cdot x \cdot x$.

There are a few different ways to read, or say, exponential numbers. Look at the example below.

$$7^5$$

This term is read as, "Seven to the fifth power." Some people shorten it by saying, "Seven to the fifth," and others will say, "Seven to the power of five." Any of these phrases are correct.

Now we will attempt to *add* some terms with exponents. Look at this next problem.

$$4^3 + 4^3 =$$

We already said that $4^3 = 64$, so that problem is saying 64 + 64. That's easy to solve, so I'll make it a little more difficult. Look at this next problem.

$$x^3 + x^3$$

Since we don't know the value of x, we can't just add them together like we did with $4^3 + 4^3$. So, let me ask you this question. Is 64 + 64 the same thing as 64 x 2? Yes, it is. So, it should be easy for you to believe that $x^3 + x^3$ is the same thing as $x^3 \cdot 2$. In algebra this is written as $2x^3$ because we like to see the coefficient in front of the variables.

$$x^3 + x^3 = 2x^3$$

To make this even clearer, I will put the invisible coefficients in front of x^3.

$$1x^3 + 1x^3 = 2x^3$$

Do you see how this problem is basically 1 + 1 = 2? As long as the terms you are adding up are **like terms** you can just add them together and put the new coefficient in front.

Can you solve this next problem?
$$x^3 + x^3 + x^3 =$$

How many x^3 do you see? I see 3, so the answer is $3x^3$.

We will look at another one.
$$3a^4 + 2a^4 =$$

Can you guess how to add these two terms together? Look at the two different coefficients of a^4. They are 3 and 2. Think of a^4 as...umm...a type of apple! Now the problem reads, "3 apples plus 2 apples," which of course equals 5 apples.

$$3a^4 + 2a^4 = 5a^4$$

However, we cannot add apples and oranges. Look at this next addition problem.

$$x^3 + a^4 =$$

We cannot add these two terms together. If "x^3" stands for oranges and "a^4" stands for apples, how can you add these together? Oranpples? No, that's not right. You cannot add these two together because they are not like terms.

The rule to be learned here is that you can only add *like terms*. And they must be exactly alike too. The coefficients don't need to be the same, just the exponents and variables. You cannot add these two terms either.

$$5a^2 + 5a^3 =$$

These cannot be added together because they are not like terms. An a^2 apple is completely different than an a^3 apple; everyone knows that, so you cannot add these terms together.

Subtraction of exponents works the same way. You can only subtract like terms. Can you solve this one?

$$4xy^2 - 2xy^2 =$$

Since these two terms are alike, we can just subtract 4 - 2 = 2, so the answer is $2xy^2$.

Can you subtract these terms?

$$7m^3 - 4m^2 =$$

No, you cannot. These are not like terms. You cannot simplify this problem any further. To *simplify* a problem means to do as much math as you can to make it smaller, simpler. Try some problems on your own by completing the next worksheet.

Name_____ Date _____

Worksheet 6

Add or subtract like terms.

1. $m^2 + m^2 =$
2. $2a^3 + 3a^3 =$
3. $4a^2b + 5a^2b =$
4. $6y^4 + 3y =$
5. $a^3 + a^3 + a^3 =$
6. $3a^5 + 2a^5 + a^5 =$
7. $7in^2 + 3in^2 =$
8. $9ab^2 + 7a^2 =$
9. $a^4 + 3a^4 =$
10. $mn + 3mn^2 + 2mn^2 =$

11. $5ab^3 - ab^3 =$
12. $9x^3 - 3x^3 =$
13. $4y^2 - 2y^3 =$
14. $15xy^4 - 12xy^4 =$
15. $11a^5 - 9a^6 =$
16. $12mn^2 - mn^2 =$
17. $10in^3 - 5in^3 =$
18. $3\pi r^2 - 2\pi r^2 =$
19. $3xy^2 - 3x^2y =$
20. $15y^2 - 5y^2 - 2y^4 =$

LESSON 7: SIMPLIFYING EXPRESSIONS

Next, we will combine several terms into one big, long problem. The math is the same as before; just look for like terms.

$$2xy^2 + 3xy^2 + 4a^2 - ab^3 + 4xy^2 - 10ab^3 =$$

To solve this problem, look for any like terms. I see a few terms with the "last name" of xy^2, so let's pull those out of the problem, along with their negative or positive signs.

$$2xy^2 + 3xy^2 + 4a^2 - ab^3 + 4xy^2 - 10ab^3 =$$

$$+2xy^2 \quad +3xy^2 \quad +4xy^2 = 9xy^2$$

Remember, if there is no sign in front of a number, it is considered to be positive. Think of "xy^2" as a type of car. How many xy^2 cars do we have? Let's do the math $2 + 3 + 4 = 9$, so the answer is $9xy^2$. Here is what the problem looks like now that we have combined all the xy^2 terms and brought down the other terms.

$$9xy^2 + 4a^2 - ab^3 - 10ab^3 =$$

I see some more like terms; ab^3 shows up twice in this problem. Let's take them out and deal with them separately.

$$9xy^2 + 4a^2 - ab^3 - 10ab^3 =$$

$$- ab^3 \quad - 10ab^3 =$$

Remember, ab^3 is the same thing as $1ab^3$. I'll rewrite that problem: $-1ab^3 - 10ab^3$. This is where your negative and positive skills come in handy. Do you know the answer to $-1 - 10 = ?$ It is -11, so when you combine the like terms above you get $-11ab^3$. Here is how the whole problem looks now that we have combined like terms.

$$9xy^2 + 4a^2 - 11ab^3 =$$

Do you see any more like terms? I don't, this problem is as simplified as it's going to get.

Sometimes you can get a little "googley-eyed" trying to find like terms in a long expression. So, I'll tell you a trick to make it easier. Look at the expression below.

$$3xyz^3 + 8yz - 4x^2 + 5xyz^3 - x^2 + 2a^4 - 3yz$$

Start by rearranging those terms in the order of their exponents. We will start with the highest exponent. Let's see… the highest exponent is a 4. We will rearrange these terms, so all of the ones with an exponent of 4 will go first. Next, we will list all terms with an exponent of 3. Make sure you move the signs along with each term. As you rewrite each term, cross off that term in the original problem. I started doing this below.

$$\cancel{3xyz^3} + 8yz - 4x^2 + \cancel{5xyz^3} - x^2 + \cancel{2a^4} - 3yz$$

$$+ 2a^4 + 3xyz^3 + 5xyz^3$$

Next, list all of the terms with an exponent of 2 and then the terms with an exponent of 1.

$$\cancel{3xyz^3} + \cancel{8yz} - \cancel{4x^2} + \cancel{5xyz^3} - \cancel{x^2} + \cancel{2a^4} - \cancel{3yz}$$

$$+ 2a^4 + \underbrace{3xyz^3 + 5xyz^3}_{+ 8xyz^3} \underbrace{- 4x^2 - x^2}_{- 5x^2} + \underbrace{8yz - 3yz}_{+ 5yz}$$

$$2a^4$$

Now it is much easier to see the like terms. The answer is:

$$2a^4 + 8xyz^3 - 5x^2 + 5yz$$

If you understand how to add and subtract terms with exponents, complete the next worksheet. If not, go back and read this lesson again.

Name_____ Date _____

Worksheet 7

Simplify the following.

1. $x + x =$

2. $2x^2 + 4x^2 =$

3. $3xy^4 + 8xy =$

4. $x^3 + 2x^3 =$

5. $ab^3 + ab^3 + 2ab^3 =$

6. $5 + 6m =$

7. $9x^4y^4 - 7x^4y^4 =$

8. $rs^3 + rs^4 =$

9. $7abc^8 + 8abc^8 =$

10. $12x^5y^3 - 6x^5y^3 =$

11. $7x^3 + 4x^3 - 1x^3 =$

12. $5xy^2 + 2xy^2 + 3a^2 - a^2 =$

13. $12ab^4 - 6ab^1 + 2ab^4 =$

14. $5x^2 + 2ab^3 - ab^3 + 3x^2 =$

15. $ax^5 + 3ax^5 - 2ax^5 =$

16. $3qrs^3 + 6qrs^3 + 4qrs^2 - qrs^2 =$

17. $12xy^2 - 11xy^3 + 10xy^3 - 3a^2 =$

18. $15abc^2 - 2abc^2 + 4abc^2 - abc^2 =$

19. $4xyz^4 + 3xy^4z - x^4yz + xyz^4 =$

20. $28mn^{10} - 14mn^{10} + 2x^2y^4z^5 + 17x^2y^4z^5 - 3mn^{10} + x^3y^4z^6 =$

LESSON 8: MULTIPLYING WITH EXPONENTS

If you feel confident adding and subtracting terms with exponents, you are ready to learn how to multiply and divide with them too. Take a look at this multiplication problem.

$$y^2 \cdot y^4 =$$

What does this problem mean? The exponents tell us how many times to use "y" in a multiplication problem.

$$y^2 \cdot y^4 =$$
(twice)   (four times)
$$y \cdot y \quad \cdot \quad y \cdot y \cdot y \cdot y$$

Squish this problem together and you get y · y · y · y · y · y. Multiplying "y" by itself 6 times, is the same thing as y^6. So, it is easy to get the answer; just add the exponents together.

$$y^2 \cdot y^4 = y^{2+4} \quad \text{or } y^6$$

Let's try another one.

$$x^2 \cdot x^5 \cdot x^4 =$$

Since the bases are the same, you can just add up the exponents 2 + 5 + 4 = 11.

$$x^2 \cdot x^5 \cdot x^4 = x^{11}$$

When the bases are not the same, as in this next problem, it's a little different - but just as easy.

$$y^2 \cdot x^2 =$$

You can't add up the exponents when the bases are different because y and x represent two different numbers. Instead, we just squish them together. When two letters are next to each other, it already means they are being multiplied. So, just remove the dot, put the variables in alphabetical order, and squish them together.

$$y^2 \cdot x^2 = x^2 y^2$$

Another thing you should realize is that when a number or letter doesn't have an exponent, it is considered to have an exponent of 1; another invisible 1!
For example, look at this next problem.

$$a \cdot a^2 =$$

That is the same thing as $a^1 \cdot a^2$, so just add up the exponents and the answer is a^3. Or simply look at that problem and ask yourself, "How many "a's" are being multiplied together?" I see 3. That is the exponent for the answer.

Now let's mix it up a little bit. I will throw in some terms with different bases. See if you can simplify this problem.

$$y^2 \cdot y^4 \cdot x^2 y =$$

How many y's are being multiplied? How many x's are being multiplied? The first term has 2 y's, the second term has 4 y's, and the third term has 1 y and 2 x's. Squish them all together, in alphabetical order, with the new exponents.

$$y^2 \cdot y^4 \cdot x^2 y = x^2 y^7$$

Now let's throw in some numbers too. Just multiply the numbers as you normally would. Tally up the exponents and put them above the appropriate base.

$$3x^2 y^2 \cdot 4xy =$$

Can you solve this one in your head? Multiply the numbers first, 3 x 4 = 12.
How many x's are there to be multiplied? How many y's are there to be multiplied? The answer is $12x^3 y^3$.

Try some on your own, by completing the next worksheet.

Name_____ Date _____

Worksheet 8

Multiply the following.

1. $x \cdot x^3 =$

2. $xy \cdot xyz =$

3. $a^4 \cdot ab \cdot b^3 =$

4. $2x \cdot 4x^2y =$

5. $3x^5y^2z^4 \cdot 7xyz =$

6. $8ab \cdot 6ab \cdot ab =$

7. $7y \cdot 2x =$

8. $3a \cdot 4a =$

9. $x^7 \cdot 2x^3 =$

10. $4mn^{10} \cdot -6mn^2 =$

11. $-1/2a(3b) =$

12. $-3b(4x^2) =$

13. $\frac{5}{6}xy \cdot \frac{1}{3}y^2 =$

14. $-5a^2 \times -6a^3 =$

15. $-a(-a^4) =$

16. $(a)^2 =$

17. $3^2 \times 3^2 =$

18. $-b^2(-c^3) =$

19. $-\frac{1}{8}a\left(\frac{3}{4}c\right) =$

20. $a^3b^4c^3 \cdot abc^2 =$

LESSON 9: DIVIDING WITH EXPONENTS

By now you should be aware that the opposite of multiplication is division. By that I mean, pick any number and multiply it by 5. Take that answer, divide it by 5, and you will be back to the number you picked. That's because division is the opposite of multiplication. You could say the same thing about addition and subtraction; they are opposite each other. With that in mind, look at this multiplication problem.

$$x^5 \cdot x^2 = x^7$$

You already know how to figure out that problem; just add the exponents like we did earlier. Now let's do the opposite operation. We will *divide* x^7 by x^5. Since dividing is the opposite, we solve this problem the opposite way. Instead of adding the exponents, like we did with multiplication, we will subtract the exponents. Look at the problem below. Make sure you are reading this problem from top to bottom as, "x to the seventh, divided by x to the fifth."

$$\frac{x^7}{x^5} =$$

To solve this type of division problem, we SUBTRACT the exponents; the opposite of how we multiplied. Subtract the numerator's exponent from the denominator's exponent.

$$\frac{x^7}{x^5} = x^2$$

Try one yourself; it's easy. Just remember the rule: to divide exponents, subtract!

$$\frac{y^4}{y^2} =$$

What is 4 - 2? What is the answer to the problem above? That's right, y^2. Look at this next problem.

$$\frac{m^3}{m} =$$

How do you solve this one? Remember, even if you don't see an exponent, there is still an invisible 1. I'll rewrite it that way.

$$\frac{m^3}{m^1} =$$

The problem is now as easy as 3 – 1 = 2, the answer is m^2. Now let's throw in some numbers. Look at the problem below. Divide the numbers as usual and then subtract the exponents.

$$\frac{24a^5}{6a^3} =$$

$24 \div 6 = 4$
$a^5 \div a^3 = a^{5-3} = a^2$
The answer is $4a^2$.

It's really just that easy. But for those of you who want to know "why," take a look at this example. I will start off easy with something you already know. You know that $24 \div 6 = 4$, right? I will rewrite that same problem.

$$\frac{24}{6} = 4$$

Now I will write that same problem again, except this time I will write the FACTORS of 24 and 6, instead. It will mean the same thing.

$$\frac{6 \cdot 4}{6 \cdot 1} = \frac{24}{6}$$

Do you see how these two problems are the same? Both numerators equal each other and so do the denominators. That makes them equal.
Watch what happens when I turn the two common factors a shade of gray.

$$\frac{6 \cdot 4}{6 \cdot 1}$$

If I got rid of those two common factors, I would be left with $\frac{4}{1}$, which is the answer! Let's try that again with some different factors of 24 and 6 to see if it will always work like that. Again, I will color the COMMON factors a shade of gray. Remember, the two problems below mean $24 \div 6$.

$$\frac{3 \cdot 8}{3 \cdot 2} \qquad \frac{2 \cdot 12}{2 \cdot 3}$$

Every time you get rid of the common factors, you are left with a fraction equal to 4.

Now let's try that same trick with variables and exponents.

$$\frac{24a^5}{6a^3}$$

Start by stretching out each term into as many factors as possible. Our goal is to cross out the common factors. So, as we are deciding which factors of 24 and 6 to use, let's select as many similar ones as possible.

$$\frac{24a^5}{6a^3} = \frac{4 \cdot 2 \cdot 3 \cdot a \cdot a \cdot a \cdot a \cdot a}{2 \cdot 3 \cdot a \cdot a \cdot a}$$

Do you see how these two division problems are the same? If not, multiply all the factors in the numerator and then multiply all the factors in the denominator. You will end up with $\frac{24a^5}{6a^3}$.

We want to get rid of as many COMMON factors as we can and then see what is left over. Remember, this is the long way to solve this problem. I prefer the short cut method I told you about earlier, but this is the behind-the-scenes look at WHY we can just subtract exponents in a division problem.

$$\frac{4 \cdot 2 \cdot 3 \cdot a \cdot a \cdot a \cdot a \cdot a}{2 \cdot 3 \cdot a \cdot a \cdot a}$$

Do you see any common factors? I will turn all the common factors a shade of gray. Now it is easy to see them.

$$\frac{4 \cdot 2 \cdot 3 \cdot a \cdot a \cdot a \cdot a \cdot a}{2 \cdot 3 \cdot a \cdot a \cdot a} = \frac{4a^2}{1}$$

I'll explain this "1" in a moment.

If I get rid of all the common factors, then I will be left with $\frac{4aa}{1}$ which equals $4a^2$.

You may be wondering why I put a number 1 underneath the $4a^2$. This "trick" of dropping the common factors is actually called *dividing out the common factors*. So, we didn't really just "get rid of" the common factors, we actually divided each one by itself. Let me explain. I'll bring back our last example.

$$\frac{4 \cdot 2 \cdot 3 \cdot a \cdot a \cdot a \cdot a \cdot a}{2 \cdot 3 \cdot a \cdot a \cdot a}$$

To "get rid of" the 2s, I have to divide each one by 2. Since $2 \div 2 = 1$, both of those 2's turn into a 1. Now let's get rid of the 3s. I will divide each one by 3. Since $3 \div 3 = 1$, they are now both 1.

$$\frac{4 \cdot \overset{1}{\cancel{2}} \cdot \overset{1}{\cancel{3}} \cdot a \cdot a \cdot a \cdot a \cdot a}{\underset{1}{\cancel{2}} \cdot \underset{1}{\cancel{3}} \cdot a \cdot a \cdot a}$$

It's the same thing with the variables. $a \div a = 1$ so, they turn into 1 as well. Now if you were to multiply all the terms together, you would end up with a 1 in the denominator.

$$\frac{4 \cdot \cancel{2} \cdot \cancel{3} \cdot \cancel{a} \cdot \cancel{a} \cdot \cancel{a} \cdot a \cdot a}{\cancel{2} \cdot \cancel{3} \cdot \cancel{a} \cdot \cancel{a} \cdot \cancel{a}} = \frac{4a^2}{1}$$

I don't know about you, but I prefer the shortcut. Let's take another look at that shortcut.

$$\frac{24a^5}{6a^3}$$

The first step is to divide the coefficients; $24 \div 6 = 4$. The second step is to subtract the exponent; $5 - 3 = 2$.

$$\frac{24a^5}{6a^3} = 4a^2$$

Let's try another one.

$$\frac{35x^4}{5x^2}$$

The first step is to divide the coefficients; $35 \div 5 = 7$. The second step is to subtract the exponents; $4 - 2 = 2$. If I were to use factors to solve this one, then it would look like this:

$$\frac{7 \cdot 5 \cdot x \cdot x \cdot x \cdot x}{5 \cdot x \cdot x} = 7x^2$$

Let's try a few more together.

$$\frac{64a^6b^5c^3}{8abc}$$

Do you know the first step? That's right, divide the coefficients; $64 \div 8 = 8$. Next, we need to subtract the exponents. Let's start with the variable "a." How many times does the variable "a" show up as a factor in the numerator? It shows up 6 times. How many times does it show up in the denominator? Just 1 time, so let's subtract: $6 - 1 = 5$.

So far, our answer is $8a^5$. Let's move onto the variable "b." There are 5 in the numerator and 1 in the denominator. That is simple subtraction; $5 - 1 = 4$. Now our answer is $8a^5b^4$. All that is left is the variable "c." I will subtract again, and our final answer is $8a^5b^4c^2$. Let's try one more together. I'll make this one really tough.

$$\frac{\frac{3}{8}x^3y^4z}{\frac{1}{2}x^3y^2}$$

Do you know where to start? Divide the coefficients. This one looks complicated, but it's just a matter of dividing fractions. Do you remember how to divide fractions? I'll rewrite it, so it looks easier to solve. The problem reads, "Three eighths divided by one half," so I'll write that. Get the reciprocal of this fraction and then multiply straight across.

$$\frac{3}{8} \div \frac{1}{2} =$$

$$\frac{3}{8} \times \frac{2}{1} = \frac{6}{8} \text{ reduces to } \frac{3}{4}$$

That is the first part of our answer. Next, we will subtract the exponents. The number of x factors are the same in the numerator and denominator, so color them gray and make them go away.

$$\frac{\frac{3}{8}x^3y^4z}{\frac{1}{2}x^3y^2}$$

Next, subtract the exponents of the "y" variable; 4 − 2 = 2.

$$\frac{\frac{3}{8}x^3y^4z}{\frac{1}{2}x^3y^2} = \frac{3}{4}y^2$$

What's left? There is a "z" variable in the numerator, but not in the denominator. No problem, there is nothing to subtract, so just leave it there. Here is the final answer.

$$\frac{\frac{3}{8}x^3y^4z}{\frac{1}{2}x^3y^2} = \frac{3}{4}y^2z$$

Try some on your own, by completing the next worksheet.

Name_____ Date _____

Worksheet 9

1. $x^5 \div x^3 =$

2. $12y^6 \div 2y^2 =$

3. $\dfrac{a^7}{a^4} =$

4. $\dfrac{9y^9}{3y^4} =$

5. $\dfrac{2}{7}a^4 \div \dfrac{1}{3}a^3 =$

6. $\dfrac{36m^5n^4}{9m^2n} =$

7. $\dfrac{42x^4y^6z^3}{7x^2y^3z} =$

8. $\dfrac{4ab^2}{2a} =$

9. $\dfrac{-8b^2}{b^2} =$

10. $\dfrac{x^2y^2}{xy} =$

11. $\dfrac{\frac{2}{5}a^2}{\frac{1}{3}a} =$ (divide the fractions)

12. $\dfrac{-9x^4y^2}{-3x^2y} =$

LESSON 10: DIVIDING TERMS WITH EXPONENTS

The problems on that last worksheet worked out nicely because every problem had the larger exponent on top. Look at this problem, it's not so simple.

$$\frac{a}{a^3}$$

> It is simple...just subtract in the opposite direction, the bottom exponent from the top.

OK, maybe it is simple. You just subtract in the opposite direction, but that does not mean that the answer is a^2. The answer is $\frac{1}{a^2}$. Let me explain. I will separate each term into factors.

$$\frac{a}{a^3} = \frac{a}{a\,a\,a}$$

Every time you see the same factor on top and bottom, you can cross them out and turn them into the number 1, because any number divided by itself, is always 1.

$$\frac{a}{a^3} = \frac{\cancel{a}^1}{\cancel{a}_1\,a\,a}$$

Now that you have turned all the common factors into 1's, re-multiply the left-over factors.

$$\frac{1}{1\,a\,a} = \frac{1}{a^2}$$

$1 \cdot a \cdot a = a^2$

Let's try another one. Can you divide the following?

$$\frac{y^2}{y^3} =$$

To make it easier to visualize, I will stretch out each term into factors.

$$\frac{y^2}{y^3} = \frac{y\,y}{y\,y\,y}$$

Every time there is the same factor on top and bottom, cross them out and turn them into the number 1.

$$\frac{y^2}{y^3} = \frac{\cancel{y}\cancel{y}}{\cancel{y}\cancel{y}y}$$

Now we look at what is left over. On the top of the fraction, we have 1 x 1 left-over, that's just a 1. The denominator has 1 x 1 x y; that equals just y. So, do you know the answer to this problem yet?

$$\frac{y^2}{y^3} = \frac{1}{y}$$

Next, I will add some numbers to our division problem; just to make it more difficult. Try this one.

$$\frac{16ab^2}{4ab^3} =$$

Start by stretching out each term into their factors.

$$\frac{16ab^2}{4ab^3} = \frac{4 \; 4 \; a \; b \; b}{4 \; a \; b \; b \; b}$$

Cross out all the matching sets of factors and turn each one into a 1.

$$\frac{16ab^2}{4ab^3} = \frac{\cancel{4}4\cancel{a}\cancel{b}\cancel{b}}{\cancel{4}\cancel{a}\cancel{b}\cancel{b}b}$$

Multiply all the factors that are left over.

$$\frac{1 \cdot 4 \cdot 1 \cdot 1 \cdot 1}{1 \cdot 1 \cdot 1 \cdot 1 \cdot b} = \frac{4}{b}$$

Look back at that last problem and try to answer it in your head. First divide the coefficients. Then notice that the "a's" will cancel each other out and that there will be 1 "b" left over in the denominator.

Try to solve this one in your head.

$$\frac{25x^2}{5x^3} =$$

Start by dividing the coefficients, 25 divided by 5. The answer is 5. Do you think that should be written as...

$$\frac{1}{5} \quad or \quad \frac{5}{1}$$

This first fraction is one fifth. The second fraction means 5, so when you divide the coefficients, the answer 5, belongs in the numerator. Next, look at what is left over.

$$\frac{x^2}{x^3} =$$

Picture in your mind 2 x's on the top and 3 x's in the denominator. Imagine if you crossed out the matching sets, what would be left over? There would be just 1 x left over in the denominator. Do you see how this is just a subtraction problem? It's just in the opposite direction, the denominator's exponent minus the numerator's exponent.

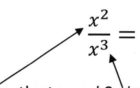

3 - 2 = 1

$$\frac{x^2}{x^3} - \frac{1}{x}$$

Don't forget about our coefficient, $\frac{5}{1}$. The entire answer is $\frac{5}{x}$.

$$\frac{25x^2}{5x^3} = \frac{5}{x} \text{ because } \frac{\cancel{25}\ 5\ \cancel{x}\ \cancel{x}}{\cancel{5}\ x\ \cancel{x}\ \cancel{x}}$$

Let's try one more together. Try to solve this next one in your head.

$$\frac{12ab^3}{3ab^6}$$

First, divide the coefficients, $12 \div 3 = 4$.

Does the answer 4 belong in the numerator or denominator?

What will you do with the "a's"?

How many "b's" are left over?

Do the left-over b's belong in the numerator or the denominator?

The 4 belongs in the numerator. The a's cancel each other out, so they are gone. The remaining 3 b's belong in the denominator.

$$\frac{12ab^3}{3ab^6} = \frac{4}{b^3}$$

Here is that same problem written with all the factors showing.

$$\frac{12ab^3}{3ab^6} = \frac{\cancel{3}\,4\,\cancel{a}\,\cancel{b}\,\cancel{b}\,\cancel{b}}{\cancel{3}\,\cancel{a}\,\cancel{b}\,\cancel{b}\,\cancel{b}\,b\,b\,b} = \frac{4}{b^3}$$

I have crossed out all identical factors that appear in both the numerator (top) and the denominator (bottom). That leaves me with just the factors that make up the answer. Try some on your own.

Name_____ Date _____

Worksheet 10

1. $\dfrac{a^2}{a^4} =$

2. $\dfrac{18y^3}{3y^6} =$

3. $a^2 \div a^6 =$ (Picture this one as $\dfrac{a^2}{a^6}$)

4. $48x^2 \div 8x^3 =$

5. $\dfrac{49mn^4}{7m^2n^6} =$

6. $\dfrac{36y^6z}{6x^2y^3z^3} =$

7. $42y^5 \div 6y^7 =$

8. $\dfrac{24ab^2}{4a^3} =$

9. $\dfrac{-9cd^2}{3c^2d^5} =$

10. $\dfrac{-16x^3y^2}{-4xy^6} =$

11. $\dfrac{\frac{3}{7}a^2b}{\frac{2}{3}ab^3} =$ (divide the fractions)

12. $\dfrac{-32x^2y^2z^2}{4x^2y^7z} =$

LESSON 11: DIVIDING EXPRESSIONS BY A SINGLE TERM

Now that you can divide single terms with exponents, we are going to divide several terms with exponents. Look at the next problem.

$$\frac{6x^2y^2 + 8xy^2 + 14x^2y}{2xy}$$

This bar means "divided by"

The first rule about dividing terms is that the denominator is dividing EACH TERM SEPARATELY. So, that division problem above is the same thing as:

$$\frac{6x^2y^2}{2xy} + \frac{8xy^2}{2xy} + \frac{14x^2y}{2xy}$$

Solve each one separately. The first term becomes 3xy. Actually, it is $\frac{3xy}{1}$, but any number over 1 is that number, so we just say 3xy. The second term becomes 4y (over 1), and the last term is 7x. Write down all those answers and be sure to put the plus signs back too.

$$\frac{6x^2y^2}{2xy} + \frac{8xy^2}{2xy} + \frac{14x^2y}{2xy}$$

$$3xy + 4y + 7x$$

Pretty easy, isn't it? Let's try another one.

$$\frac{4x^2 + 8xy^2 - 24y^2}{4x^4}$$

Can you solve this one on your own? Can you solve it in your head and just write down the answer? Give it a try. Remember to divide EACH term by the denominator.

$$\frac{4x^2 + 8xy^2 - 24y^2}{4x^4} = \frac{4x^2}{4x^4} + \frac{8xy^2}{4x^4} - \frac{24y^2}{4x^4}$$

Do the math.

$$\frac{4x^2}{4x^4} = \frac{1}{x^2}$$

$$\frac{8xy^2}{4x^4} = \frac{2y^2}{x^3}$$

$$\frac{24y^2}{4x^4} = \frac{6y^2}{x^4}$$

The final answer is $\frac{1}{x^2} + \frac{2y^2}{x^3} - \frac{6y^2}{x^4}$

But wait! There is another way to solve that last problem. Take another look. When I look at these coefficients, I see four numbers with a common factor.

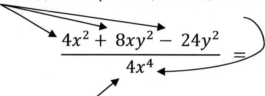

That common factor is 4, so let's do some canceling, like we did before. I will divide each term by 4.

$$\frac{\overset{1}{\cancel{4}}x^2 + \overset{2}{\cancel{8}}xy^2 - \overset{6}{\cancel{24}}y^2}{\underset{1}{\cancel{4}}x^4} = \frac{x^2 + 2xy^2 - 6y^2}{x^4}$$

I can't cancel any of the "x's" because not every term has an x as a factor. I can only cancel the common factors of EVERY term, **including the denominator**.

Now look at the two different answers I got from the same problem.

Answer #1

$$\frac{1}{x^2} + \frac{2y^2}{x^3} - \frac{6y^2}{x^4}$$

Answer #2

$$\frac{x^2 + 2xy^2 - 6y^2}{x^4}$$

How can both of these be the right answer? Well, I will show you, and it's pretty cool, just watch. In answer #2, the denominator is x^4. In answer #1, we have three different denominators. We need to make all three denominators the same. We want all three denominators to be x^4, just like the denominator in answer #2.

$$\frac{1}{x^2} + \frac{2y^2}{x^3} - \frac{6y^2}{x^4}$$

We will start with the first fraction in answer #1. Think about that...what will it take to turn the denominator into x^4? Multiply it by x^2; that will turn x^2 into x^4. And of course, whatever you do to the denominator you must do to the numerator.

$$\frac{1}{x^2} \cdot \frac{x^2}{x^2} = \frac{1x^2}{x^4}$$

Original fraction   New fraction
Number used to get a common denominator.

Next, we will tackle the second fraction in answer #1, $\frac{2y^2}{x^3}$. What will it take to turn x^3 into x^4? It needs to be multiplied by one more x.

$$x^3 \cdot x = x^4$$

And whatever you do to the denominator, you must do to the numerator.

$$\frac{2y^2}{x^3} \cdot \frac{x}{x} = \frac{2xy^2}{x^4}$$

Original fraction   New fraction
Number used to get a common denominator.

And finally, the third fraction in answer #1, $-\frac{6y^2}{x^4}$. This fraction already has a denominator of x^4, so we don't have to change this one. Now, put the three new fractions together.

$$\frac{1x^2}{x^4} + \frac{2xy^2}{x^4} - \frac{6y^2}{x^4}$$

Since all the denominators are the same, you should rewrite that fraction with just one denominator. Now it looks just like answer #2.

$$\frac{1x^2 + 2xy^2 - 6y^2}{x^4}$$

We are going to pause here for a moment. Take a second look at the third fraction in answer #1. The negative sign can be confusing. Sometimes the sign is in the numerator, sometimes it is in the denominator, and sometimes it is in front of the fraction...where does it go? I will explain by using math you already understand.

$$-56 \div 8 = -7 \qquad -56 \div -8 = 7 \qquad 56 \div -8 = -7$$

These problems can be rewritten like this...

$$\frac{-56}{8} = -7 \qquad \frac{-56}{-8} = 7 \qquad \frac{56}{-8} = -7 \qquad -\frac{56}{8} = -7$$

Notice how it only takes one negative sign to make the answer negative. When both signs are negative, the answer is positive. When you look at negative fractions as division problems, it's easy to find the answer.

Keep in mind, there are 3 different places that a negative sign can show up in a fraction: the numerator, the denominator and in front of the fraction. You need to use your negative/positive skills to mathematically figure out each one. For example, look at this next problem.

$$-\frac{-21}{-3} =$$

When you do the math, $-21 \div -3$ you get positive 7. BUT, there is a negative sign in front of the fraction, so now it is negative 7. You could look at that same problem like this...

$$-\left(\frac{-21}{-3}\right) = -7$$

The Order of Operations tells us to solve the math inside the parentheses first. Then we apply the negative sign to that answer.

$$-\left(\frac{-21}{-3}\right) =$$
$$\downarrow$$
$$-(7) = -7$$

Look at this problem.

$$-\left(\frac{16}{-4}\right) =$$

Solve the math inside the parentheses first, $16 \div -4 = -4$. Now look at the sign in front of the parentheses. This negative sign turns our –4 into a negative, negative 4...and that means positive 4. Even if the negative sign had been in the numerator, the answer would still be +4.

$$-\left(\frac{-16}{4}\right) = 4$$

The lesson to be learned here is to use your negative/positive skills to understand a fraction with a negative sign. Look at this one. What do you get for an answer?

$$-\frac{12}{4} =$$

First solve the math, $12 \div 4 = 3$. Then apply the negative sign. The answer is -3. Now let's get back to dividing expressions.

Let's try to solve another "expression divided by a single term" together. Look for a common factor in every term.

$$\frac{6a + 9a - 12ab}{3ab^2} =$$

All of the coefficients have a common factor of 3. Look at the variables in each term. Do you see any common factors there? Each term has the letter "a" as a factor. That means our common factor is 3a. We will divide each term by 3a.

$6a \div 3a = 2$

$9a \div 3a = 3$

$-12ab \div 3a = -4b$

$3ab^2 \div 3a = b^2$

$$\frac{\overset{2}{\cancel{6a}}+\overset{3}{\cancel{9a}}-\overset{4}{\cancel{12ab}}}{\underset{1}{\cancel{3ab}}^2} =$$

Can you write the answer to this problem now?

$$\frac{2 + 3 - 4b}{b^2} =$$

The only math we can do is add 2 + 3, so this is the final answer.

$$\frac{2 + 3 - 4b}{b^2} = \frac{5 - 4b}{b^2}$$

Try to figure this one out on your own.

$$\frac{6a - 36b^2 + 12a^2b}{-6ab} =$$

Can you find a common factor? Cross out the coefficients and divide by that common factor. Is there a variable in the common factor? This one is a little bit challenging, so double check your negative/positive work. After you get your answer, compare it to my answer.

What did you get for a common factor? Did you go with negative 6 or positive 6? Surprisingly, they will both work, so which one should you use? If all the terms had been negative, it would make more sense to go with a negative common factor. Or if all the terms were positive, then of course we would go with a positive number. But in this problem, there are two positive terms and two negative terms. We will try both common factors and then compare the results.

First, we will divide each term by positive 6.

$$\frac{\overset{1}{\cancel{6}}a - \overset{-6}{\cancel{36}}b^2 + \overset{2}{\cancel{12}}a^2 b}{\underset{-1}{\cancel{-6}ab}} =$$

$$\frac{a - 6b^2 + 2a^2 b}{-ab} =$$

That gave us a negative denominator, which is kind of weird. Let's divide each term by NEGATIVE six and see what we get.

$$\frac{\overset{-1}{\cancel{6}}a - \overset{6}{\cancel{36}}b^2 + \overset{-2}{\cancel{12}}a^2 b}{\underset{1}{\cancel{-6}ab}} =$$

$$\frac{-a + 6b^2 - 2a^2 b}{ab} =$$

Do you see how dividing by negative 6 changes every sign in the problem to the opposite sign? So which answer is correct? Well, they are actually the same number, but leaving the answer with a positive denominator is the better choice. Do you wonder how both answers can be the same number? You should. Let's assign some numbers to "a and b," so we can do the math in both answers. We'll say $a = 2$ and $b = 3$. I'll fill in both answers with numbers, instead of variables.

Are these two expressions equal?

$$\frac{a - 6b^2 + 2a^2b}{-ab} = \frac{-a + 6b^2 - 2a^2b}{ab}$$

$a = 2$ and $b = 3$

$$\frac{2 - 6(9) + 2(4 \cdot 3)}{-(2 \cdot 3)} = \frac{-2 + 6(9) - 2(4 \cdot 3)}{(2 \cdot 3)}$$

$$\frac{2 - 6(9) + 2(12)}{-6} = \frac{-2 + 6(9) - 2(12)}{6}$$

$$\frac{2 - 54 + 24}{-6} = \frac{-2 + 54 - 24}{6}$$

$$\frac{-28}{-6} = 4.\overline{66} \quad = \quad \frac{28}{6} = 4.\overline{66}$$

Do you see that? They both equal 4.66 when you do the final division because in each fraction, the numerator and denominator have the same signs, so the answer is positive either way. This is a difficult lesson, so if you are unclear on what I've done, read this lesson again, otherwise complete the next worksheet.

Name_____ Date _____

Worksheet 11

Simplify the following by dividing.

1. $\dfrac{2ab+4a+10ab^2}{2a} =$

2. $\dfrac{10x-25xy-15x^3}{5x^2} =$

3. $\dfrac{28mn+14m-49n^2}{7m^2n^2} =$

4. $\dfrac{42k-30k^2+18k^3}{-6k^3} =$

5. $\dfrac{-27y^4-9y^3-18y^2}{-3y^7} =$

6. $\dfrac{-64b-48ab+56a^2}{8a} =$

7. $\dfrac{16rs^2+12r^2s+20r^3s^3t}{4rst^2} =$

8. $\dfrac{11w^3+4w+9w^2}{2w^3} =$

9. $\dfrac{ab-ab^2-ab^3}{ab^2} =$

10. $\dfrac{11d+22ef^2+121f}{11def} =$

11. $\dfrac{6xy^2z+4xy^2z+8xy^2}{12xy^2z} =$

12. $\dfrac{-54c+45c+9c^2}{-9c^4} =$

LESSON 12: YOU CAN ONLY CANCEL FACTORS – NOT TERMS

There is one very important concept that you must, MUST understand. It is OK to cross out FACTORS in a division problem, but you CANNOT cross out terms. Factors are numbers or variables being multiplied together. For example, look at the following four terms and some of their factors.

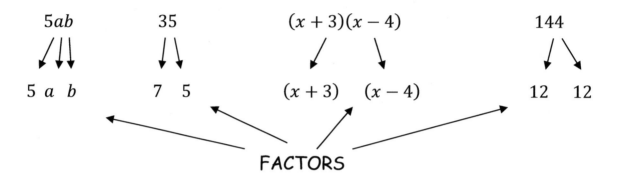

Since this is so important, let's go over the terminology one more time. I will use the number 20 for this example. Can you name some factors of 20? Let's see, what two numbers can be multiplied together to get 20? I know! 4 and 5 equal 20 when multiplied together. We will use those two factors.

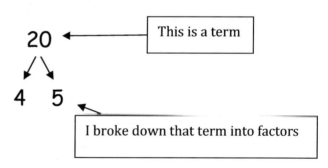

Now I will take the number 20 again, but this time I will break it down into two numbers that equal 20 when ADDED together not multiplied.

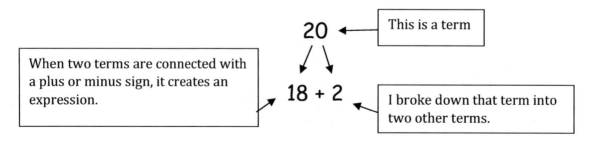

The reason this is a big deal is because in algebra you will end up with these huge division problems and the only way to solve them is by canceling out common factors. If you aren't clear on which factors to cancel, you will have a difficult time solving those types of problems. This may seem difficult at first, but you'll get the hang of it, and you will do well in algebra. This is one of the hardest parts of algebra.

Let me explain further by giving you a simple division problem and then separating it into factors.

$$\frac{6}{3} = \frac{3 \cdot 2}{3 \cdot 1} \leftarrow \text{Factors of 6 and 3}$$

I already know that 6 ÷ 3 = 2, but I'm trying to make a point, so let's cross out the common factors and solve it the long way.

$$\frac{6}{3} = \frac{\cancel{3} \cdot 2}{\cancel{3} \cdot 1} = \frac{2}{1} = 2$$

Since there is a FACTOR of 3 in both the numerator and denominator, I can cross them out and turn them both into a 1. Multiply the numbers now and you'll end up with 2, which we know is the correct answer. Now I will write the same division problem, but instead of using FACTORS of 6, I will use and expression to say "6."

$$\frac{3+3}{3} =$$

On top we have 3 + 3. These are not factors of 6. 3 + 3 is an expression consisting of two terms being added together. If I mistakenly tried to cross out the 3 from the top and bottom, I would get the wrong answer.

NOT FACTORS

WRONG! ⟶ $\frac{\cancel{3}+3}{\cancel{3}} = \frac{3}{1} = 3$

We know this is wrong because 6 ÷ 3 does not equal 3! The correct way to solve that division problem is to add the numbers together first and then divide. Or break it down into factors, if you couldn't solve it in your mind.

$$\frac{3 + 3}{3} = \frac{6}{3}$$

The rule here is YOU CAN ONLY CANCEL FACTORS!! This is important!! Remember, when it comes to canceling: Factors are Fine, Terms are Trouble! But this can be confusing, since sometimes a "term" can be full of factors. Let me explain with the next example.

$$\frac{4xy + 4x}{4}$$

This is a division problem. On the top, we have an expression, $4xy + 4x$. Let's focus on that expression for just a moment.

$$\frac{4xy + 4x}{4}$$

That expression is a term ($4xy$) plus another term ($4x$). When you connect terms together with a plus sign, you create an expression. Now let's focus on each of those terms. The first one, 4xy, means, "four times x times y." What are the factors of the term? Well, that's easy, they are 4, x, and y. Those are the "numbers" being multiplied together, so those are the factors.

How about the second term in our expression, 4x. What are the factors of 4x? Factors are the numbers being multiplied, so they must be 4 and x.

Now take a look at the number on the bottom. It is a 4. What are the factors of 4? There are two sets of factors for 4. First is 4 and 1. Second is 2 and 2. We are trying to get a COMMON factor in each term, so let's go with 4 and 1. That way we can use 4 as our common factor.

So, now that we have determined that all of these numbers and variables are indeed FACTORS, we can cancel as we see fit.

$$\frac{4xy + 4x}{4 \cdot 1}$$

We have decided that 4 is our COMMON factor (the one that each term has in common), so let's cross them out and turn them into 1's. The reason we turn them into 1's is because we are actually dividing each term by our common factor. If this number here had been a 12, our common factor would still be 4, but when

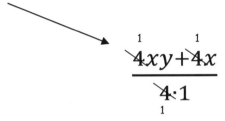

we crossed out the 12 it would turn into a 3, because we are dividing by our common factor. Now that we have gotten rid of, or divided out, our common factor let's multiply what is left over.

$$\frac{1xy + 1x}{1}$$

We can just drop all those 1's. It will mean the same thing because 1 times any number is that number and any number over 1, is that number.

$$\frac{4xy + 4x}{4} = xy + x$$

Just remember you can cancel Factors - that's Fine. But you cannot cancel a Term - that's Trouble.

Let's go over a few more examples together. Look at the next problem.

$$\frac{a+5}{a}$$

→ This is a division problem. I know it looks like a fraction. That's because a fraction IS a division problem. Remember, $\frac{1}{2}$ equals .50 because $\frac{1}{2}$ is the same thing as one divided by two. This was taught in volume III of the *Learn Math Fast System*.

Let's get back to our division problem. It is "a plus 5, divided by a."

$$\frac{a+5}{a}$$

Can we cancel the two a's? NO, NO we cannot! In this case, the "a" is a TERM, not a factor. How do I know that? Look at the expression "a + 5." Nothing is being multiplied here, so there aren't any factors. This problem is as simplified as it is going to get. We would need to know what "a" is equal to in order to go any further.

Here is another one. Can you find the common factor or factors?

$$\frac{42a}{7a}$$

I will solve this one three different ways. First, I will expand each term into their factors. So, let's see, what are some factors of 42? The first ones that come to mind are 6 and 7. I will rewrite our division problem with those factors plus the "a" variable.

$$\frac{6 \cdot 7 \cdot a}{7 \cdot a}$$

The factors of "7" are 1 and 7, but it isn't necessary to write the "1" because 1 times anything...is that thing. Now I can just cross out the common factors and then see what's left.

64

$$\frac{6\cdot\cancel{7}\cdot\cancel{a}}{\cancel{7}\cdot\cancel{a}}$$

Our common factors were 7 and a. Since they were identical on both the top and bottom, they are now all 1's. So, what's left? Multiply them.

$$\frac{6\cdot 1\cdot 1}{1\cdot 1}=\frac{6}{1}=6$$

The answer is 6. Now I will solve the same problem a different way.

$$\frac{42a}{7a}$$

As I look at this problem, I notice that 7 is a common factor of both 42 and 7. This time I will divide each coefficient by 7. That will turn the "7" into a "1" and it will turn the "42" into a "6."

$$\frac{\overset{6}{\cancel{42}}a}{\underset{1}{\cancel{7}}a}$$

Since there is an "a" on top and bottom, I can just cross out those common factors. Technically, they are both 1's now, but what's the point? Anytime you multiply a number by 1, it won't change the number, so it I'm not going to write it. It is easy to see that the answer, once again, is 6.

But there is still one way that is even faster. Look at the problem one more time. It is a division problem, so let's treat it like one. What is $42 \div 7$?

$$\frac{42a}{7a}$$

It is 6. I know that the two a's cancel each other out, so they are gone, and the answer is 6.

Let's try one more together. What are the factors of $5(x+y)$?

$$\frac{5(x+y)}{5y} \leftarrow$$

Not sure? What is being multiplied? 5 and (x + y) are being multiplied, so those are the factors. What are the factors of 5y? They are 5 and y. So then, what is our common factor?

$$\frac{5(x+y)}{5y}$$

Our common factor is 5. I will mentally cross out the 5's and see what is left.

$$\frac{(x+y)}{y}$$

Can I cross out the y's? NO, no I cannot. The "y" on the top is not being multiplied by anything, it is being ADDED to "x," so it is a term not a factor. This problem is solved.

Think you got it? Then complete the next worksheet. Not so sure yet? Read this lesson again.

Name_____ Date _____

Worksheet 12

Find a common factor, if there is one, and then simplify the following.

1. $\dfrac{2+a}{2} =$

2. $\dfrac{10x}{2x} =$

3. $\dfrac{7ab^2}{ab} =$

4. $\dfrac{x+y+z^3}{xyz^5} =$

5. $\dfrac{7+a}{a} =$

6. $\dfrac{24a}{3a} =$

7. $\dfrac{5ab+5b}{5} =$

8. $\dfrac{3(a+b)}{3} =$

9. $\dfrac{a+b}{a} =$

10. $\dfrac{4ac}{ac} =$

11. $\dfrac{(y-1)}{y} =$

LESSON 13: SQUARED AND CUBED

Now that you have learned how to work with exponents, let me tell you one last little math fact that most people don't realize.

I already told you how to read a number with an exponent. For example, 8^4 is read as, "Eight to the fourth." There are two numbers, however, that get a special name, when used as an exponent. They are the numbers 2 and 3. Whenever the exponent is the number 2, as in 6^2, it is read as "six squared." If the exponent is the number 3, as in 7^3, it is read as "seven cubed."

Let's dig a little deeper into math, so you can fully understand why they are called squared and cubed, because knowing this is often the difference between the people who go on to higher math and those who don't.

Do you remember how to find the area of a square from Geometry class? You multiply the two sides of a square to get the area, right? Let's look at the math to find the area of a square with 6-inch sides.

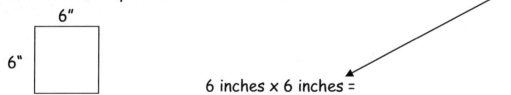

That's easy! The answer is 36 square inches. But we are going to dig even deeper into the world of math and find out *why* the answer is 36 square inches.

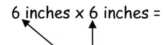

To solve this problem, you multiplied (6 x 6) and then (inches x inches) which is the same thing as $6^2 in^2$. We can solve 6^2; that's easy, 6 x 6 = 36. Can you do anymore math to inches^2? No, so the answer is $36 in^2$.

Do you see why there will always be an exponent of ^{2} in the answer to the area of a square? Because the unit, whether it's feet, miles, inches, or yards, will always be multiplied by itself, so the answer will always be "squared units." Now do you understand why the exponent 2 is read as squared? Well, keep reading and you will.

Next, we will examine the cube³. In Geometry you learned how to find the volume of a cube. You multiply all three sides, right?

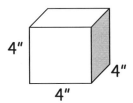

This problem is 4 inches x 4 inches x 4 inches. Since this is the same thing as 4 x 4 x 4 x inches x inches x inches, we could also say 4³inches³. We can solve 4³. That's 4 x 4 x 4 = 64, so the answer is 64inches³. This is read as sixty-four cubic inches or you could say sixty-four inches cubed, either way is correct. Do you see how we multiplied 3 numbers from our cube, we ended up with an answer that has an exponent of 3, and we call it cubed?

Here is the point I'm trying to make. The area of a **square** is found by multiplying **2** numbers. The answer will always have an exponent of **2** and will be read as **squared**. The volume of a **cube** is found by multiplying **3** numbers. The answer will always have an exponent of **3** and will be read as **cubed**. Make sense?

Earlier, we said 6" x 6" = 36in². Another way to write that is...

$$6 \text{ inches} \times 6 \text{ inches} = (6 \text{inches})^2$$

But look at that amount carefully. If you just drop the parentheses you will have 6inches², but that's not the right answer. The answer should be 36inches². The rule here is that you must apply the exponent to each factor inside the parentheses, so (6inches)² = 6²inches².

But remember, this rule only applies when there are FACTORS inside the parentheses. Look at these 4 sets of parentheses, can you tell which one does not contain factors?

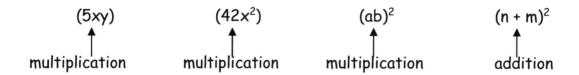

The last set of parentheses $(n + m)^2$ does not have any factors. Since there is addition inside the parentheses, you cannot square n and m individually.

When the amount inside the parentheses is an expression, as in $(5 + 3)^2$, it means $(5 + 3)(5 + 3)$. You wouldn't square each number separately because then your answer would be $(5^2 + 3^2)$, which is 25 + 9, instead of 8 × 8. In algebra, they like to write out these rules using letters to represent any number.

Here are the last two rules written out algebraically. Notice the difference between the two. The first one has a multiplication problem inside the parentheses. The other one has an addition problem inside.

$$(ab)^2 = a^2b^2 \qquad\qquad (a + b)^2 = (a + b)(a + b)$$

This rule is saying, if you square a multiplication problem inside of parentheses, you must square each FACTOR. The other rule says, if you square an expression that is inside parentheses, you square the whole thing as one amount.

I can prove these two rules are true, by looking at math we already know. Look at the examples below and read each step until you understand it. I will replace the variables "a and b" with the numbers 4 and 2, in other words, a = 4 and b = 2.

$$(ab)^2 = a^2b^2$$
$$(4 \times 2)^2 = 4^2 2^2$$
$$8^2 = 16 \times 4$$
$$64 = 64$$

That should prove to you why $(ab)^2$ is the same thing as a^2b^2. Let's look at the other rule. Again, I will make a = 4 and b = 2. Try to follow each step.

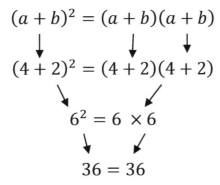

If you had done that last problem incorrectly, by squaring each of the terms inside the parentheses, you would have come up with $4^2 + 2^2$. But that equals 20 and that is the wrong answer.

Now do you understand the two new rules? Are you wondering why anybody would ever need to know these rules and why you should even care that $(ab)^2 = a^2b^2$? Well, that's a fair question and I have two answers.

Number one, when there are numbers inside of parentheses, as in $(4 + 2)^2$, it is easy to solve with math. But when there are variables inside of parentheses, as in $(a + b)^2$, you need to have a rule to know how to solve it because you can't see the numbers. That's why they give you a rule, so your answer makes sense whether there are numbers available or not.

The second reason you need to learn this rule is because at the end of this book you are going to solve HUGE mathematical equations, and I'm teaching you all the rules you are going to need to be able to solve them.

Use those two new rules to simplify the problems on worksheet 13.

Name_____ Date _____

Worksheet 13

Rewrite each value without parentheses or without an exponent. The first one is done for you.

1. $(5x)^2 = 5^2 x^2 = 25x^2$
2. $(6+2)^2 =$
3. $(xy)^3 =$
4. $(2ab)^4 =$
5. $(x+3)^2 =$
6. $(10-7)^3 =$
7. $(12-a)^2 =$
8. $(a+b)^3 =$
9. $(3y)^2 =$
10. $(4xy)^2$

*Review: Add, subtract, multiply, or divide the following.

11. $2x^4 + 4x^4 =$
12. $6mn^2 - 3mn^2 =$
13. $5xy^3 \cdot -3x^2y^2 =$
14. $\frac{24a^2b^2}{3ab} =$
15. $8x + 8x^2 =$
16. $34a^3 - 3b^3 =$
17. $9ab^4c^2 \cdot 8a^3b^3c =$
18. $\frac{4xy^2}{2xy^4} =$
19. $3xyz^3 + 4xyz^2 =$
20. $4r^2s^2t^3 \cdot 11rs^3t^2 \cdot 2rst =$

Check each answer and learn from any mistakes you may have made.

LESSON 14: SQUARE ROOTS

Earlier I said that the opposite of addition is subtraction. I also said the opposite of multiplication is division. But did you know that the opposite of exponents are roots? The most common are square roots!

By now you should know that $6^2 = 36$. Do you know the answer to this problem?

$$\sqrt{36} =$$

Of course, you do. This problem is read, "What is the square root of 36?" The square root is whatever number you squared to get 36. We squared 6 to get 36, so the answer to the problem above is 6. What is the answer to this next problem?

$$\sqrt{64} =$$

What is the square root of 64? Think to yourself, 6 x 6 = 36…that's not enough so try 7 x 7 = 49…still not enough…try 8 x 8 that equals 64! Bingo! The square root of 64 is 8.

Answer the following square root problems.

Name_____ Date _____

Worksheet 14

1. $\sqrt{16} =$

2. $\sqrt{4} =$

3. $\sqrt{49} =$

4. $\sqrt{81} =$

5. $\sqrt{25} =$

6. $\sqrt{9} =$

7. $\sqrt{100} =$

8. $\sqrt{144} =$

9. $\sqrt{121} =$

10. $\sqrt{36} =$

This should be simple for you. If you have problems with multiplication, you should read the *Learn Math Fast System Volumes I - III* before continuing.

LESSON 15: CUBE ROOTS

That last worksheet was easy, so I'll make it a little more challenging. Can you answer this next problem?

$$3^3 =$$

This problem is asking you to solve, "three cubed."

$$3 \times 3 \times 3 =$$

We can solve that, $3 \times 3 = 9$ and $9 \times 3 = 27$, so the answer is 27.

$$3^3 = 27$$

But can you solve this problem? Take a guess.

$$\sqrt[3]{27} =$$

This problem is asking you to solve the **cube root** of 27. If $3^3 = 27$, then $\sqrt[3]{27} = 3$. Try to answer this one, it's kind of tough.

$$\sqrt[3]{125} =$$

The best way to solve this problem, besides guessing a number and using a calculator, is to find factors of 125. As you are factoring, try to get 3 identical factors. Can you think of two numbers that equal 125 when multiplied together? Since 125 ends with the number 5, I know it is divisible by 5. So the question is 5 times what equals 125? I'll make a factor tree.

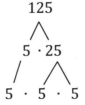

The factor tree shows that 5 x 5 x 5 = 125, so our answer is 5.

$$\sqrt[3]{125} = 5$$

Let's try another one.

$$\sqrt[3]{27} =$$

Can you find factors of 27? Keep factoring until you come up with 3 identical factors.

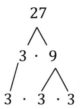

This factor tree tells us that the cube root of 27 is 3. Sometimes the math doesn't work out so perfectly. For example, let's try to get the cube root of 216, by factoring the number 216. We will cover factoring more thoroughly later in this book, but you probably learned how to draw factor trees in elementary school.

$$\sqrt[3]{216} =$$

Can you think of two numbers that equal 216 when multiplied together? Since 216 is an even number, I know that 2 must be a factor. Do you know the other factor?

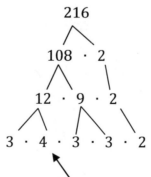

I still don't see 3 identical factors, but look a little closer. What are the factors of 4? Add those factors to our factor tree and then we'll look for identical factors.

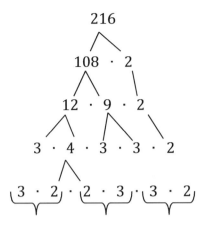

OK...I still don't see 3 identical factors, but I do see 3 groups of factors. Now can you answer this problem?

$$\sqrt[3]{216} =$$

Looking at our factor tree, I see that each group of factors are 2 x 3. That means that 6 x 6 x 6 = 216 (or 6^3 = 216), so the cube root of 216 is 6.

You may be wondering, if $\sqrt[3]{27}$ means the cube root of 27, then shouldn't we write $\sqrt[2]{64}$ to say the square root of 64? The answer is yes, you could write it that way, but the square root symbol already implies that we are looking for the "square" root. We only need to put a little number to the left, if we are looking for a root other than the square root.

One more thing you must consider. If you square a negative number, you will always get a positive answer.

$$-4^2 = 16 \quad because \quad -4 \times -4 = 16$$

And when you square a positive number, you will also get a positive answer.

$$4^2 = 16 \quad because \quad 4 \times 4 = 16$$

So then, what is the correct answer to the square root of 16?

$$\sqrt{16} = ?$$

Is it negative 4 or positive 4? Since it could be either -4 or +4, mathematicians insist that we write the answer like I did below. It is read as, "Plus or minus four."

$$\sqrt{16} = \pm 4$$

It is an annoying little symbol because if you forget to write it, some teachers will mark the answer as wrong. With that being said, ponder over this one for a moment.

$$\sqrt{-64} =$$

What is the square root of negative 64? What two identical numbers were multiplied together to get -64? Mmm...there is no such number. Writing the square root of a negative number is so ridiculous, that it isn't even considered a real number.

Now you CAN have a negative number under a *cube* root sign because if you multiply 3 negative numbers together, your answer will be negative, look below.

$$-5^3 = -125 \qquad \sqrt[3]{-125} = -5$$
$$\text{Because}$$
$$-5 \cdot -5 = 25 \quad and \quad 25 \cdot -5 = -125$$

Try some on your own, by completing the next worksheet.

Name_____ Date _____

Worksheet 15

Solve the following problems.

1. $\sqrt[3]{8} =$

2. $\sqrt[3]{64} =$

3. $\sqrt[3]{216} =$

4. $\sqrt[3]{27} =$

5. $\sqrt[3]{125} =$

6. $\sqrt[3]{-64} =$

7. $\sqrt[3]{-8} =$

8. $\sqrt[3]{-27} =$

9. $\sqrt[3]{-1000} =$

10. Bonus problem: $\sqrt[4]{81} =$

11. $\sqrt[3]{8} + \sqrt[3]{8} =$

12. $\sqrt[3]{8} \cdot \sqrt[3]{-8} =$

13. $\dfrac{\sqrt[3]{-216}}{\sqrt[3]{8}} =$

Check each answer, to make sure you fully understand this lesson.

LESSON 16: SQUARE ROOT OF A FRACTION

Take a look at this next problem. It looks scary, but it's really easy.

$$\sqrt{\frac{9}{16}}$$

This problem is read as, "the square root of nine sixteenths." This can be simplified by getting the square root of 9 and then the square root of 16. Here are the steps.

$$\sqrt{\frac{9}{16}} = \frac{\sqrt{9}}{\sqrt{16}} = \frac{3}{4}$$

You can check to see if I got it right. If the square root of $\frac{9}{16}$ is $\frac{3}{4}$, then...

$$\frac{3}{4} \cdot \frac{3}{4} \text{ must equal } \frac{9}{16}$$

And it does, so I got the right answer. Try one on your own.

$$\sqrt{\frac{4}{25}} =$$

This is simple. What is the square root of 4? What is the square root of 25? Take your answer and square it. Did you get back to $\frac{4}{25}$? Then you got the right answer. The answer is $\frac{2}{5}$.

Sometimes, you won't be able to write the square root of a number, for example $\sqrt{b}$. No big deal. When this happens, just leave the square root symbol over that number or variable.

$$\sqrt{\frac{b}{9}} = \frac{\sqrt{b}}{3} \qquad \text{or} \qquad \sqrt{\frac{5}{64}} = \frac{\sqrt{5}}{8}$$

Name_____ Date _____

Worksheet 16

1. $\sqrt{\dfrac{9}{36}} =$

2. $\sqrt{\dfrac{2}{25}} =$

3. $\sqrt{\dfrac{4}{x}} =$

4. $\sqrt{\dfrac{7}{64}} =$

5. $\sqrt{\dfrac{16}{81}} =$

6. $\sqrt{\dfrac{b}{121}} =$

7. $\sqrt{\dfrac{a}{c}} =$

8. $\sqrt{\dfrac{7}{9}} =$

9. $\sqrt{\dfrac{9}{10}} =$

10. $\sqrt{\dfrac{25}{49}} =$

Name_____ Date _____

CHAPTER 2 REVIEW TEST

Add the following.
1. $a^3 + a^3 =$
2. $5ac^2 + ac^2 =$
3. $7y^2 + 5y^2 + 2y^2 =$
4. $9x^4y^5z^6 + 7x^4y^5z^6 =$
5. $4b^3 + 2b^3 + 6y^2 =$
6. $6ab^3 + 4xy^3 + 4ab^3 + 6xy^3 =$

Subtract the following.
7. $12m^4 - 7m^4 =$
8. $-13y^3 - 8y^3 =$
9. $5r^2 - (-r^2) =$
10. $8^2 - 6^2 =$
11. $6jk^2 - m^3 - jk^2 =$
12. $-abc^4 - (abc^4) =$

Multiply the following.
13. $9a^2 \cdot 3a^2 =$
14. $xy^3 \cdot xy =$
15. $3^2 \cdot x^2 =$
16. $(7e^3f^2g)^2 =$
17. $-\frac{2}{5}x^2 \cdot \frac{2}{3}x^3 =$
18. $(-2xy)^3 =$

Divide the following.
19. $\frac{a^5}{a^2} =$
20. $\frac{a^3}{a^5} =$

21. $\frac{24x^2y^3}{4xy} =$
22. $\frac{-48m^6}{8m^{12}} =$

23. $54ab^2 \div -6ab^4 =$
24. $\frac{10ab^3 + 8a^2 + 6b}{2ab} =$

25. $\frac{32c^2 + 8a^2c^2 - 12abc^2}{4ac^2} =$
26. $\frac{-a+b-a}{ab^2} =$

Fill in the blank.
27. Canceling factors is _____, but canceling terms is _____.

Cancel, if you can.
28. $\frac{a+b}{ab}$
29. $\frac{5ab}{7ab^2}$
30. $\frac{6+a^3}{6a^3}$

82

Chapter 2 Test Continued...

31. $\dfrac{18a^2b^2c^2}{6abc} =$

32. $\dfrac{7(a+b)}{7xy} =$

33. $\dfrac{45y+18y^2+27yz}{-9yz^2} =$

34. Which one of the following terms is the only possible answer to the volume of a cube?

    8in²   9xy²   64in³   1/2in⁴   7ft²

35. Which one of the following terms is the only possible answer to the area of a square?

    5in³   4ft.   9cm²   8 miles   π

Solve the following.

36. $\sqrt{81} =$

37. $\sqrt[3]{-125} =$

38. $\sqrt{\dfrac{16}{25}} =$

39. $\sqrt[2]{25} =$

40. $\sqrt{\dfrac{9}{b}} =$

41. $\sqrt{49} - \sqrt[3]{-8} =$

42. $(\sqrt[3]{216})^2 =$

43. $\sqrt{-9} =$

44. $\dfrac{\sqrt[3]{-512}}{\sqrt{16}} =$

45. $\dfrac{a^2-4ac}{4a^2} =$

46. $\dfrac{28a^3 + \sqrt{25} - 3a^3}{5a^3} =$

47. $\dfrac{-3x^2 - 5x^3 + 9x}{-x^4} =$

48. $33x^2y^3 - 14a^2b + x^2y^3 - 9x^2 - 3a^2b + 8x^2 - (4^2x^2y^3) =$

CHAPTER 3   LAWS OF ALGEBRA

LESSON 17: LAWS OF ADDITION

Next, you will learn two very important and very simple Laws of Algebra that will help you solve huge complex algebra problems. The first Law is called the *Commutative Law of Addition*. This law is really simple; it states that if 3 + 5 = 8, so does 5 + 3. That's it! This law says it's OK to switch the two numbers around because they will still add up to the same amount. You can't say the same thing for subtraction. 5 – 3 is not the same as 3 – 5, is it? Here is the template for the Commutative Law of Addition.

$$a + b = b + a$$

Fill in "a and b" with the numbers 3 and 5.

$$3 + 5 = 5 + 3$$

This should be really obvious and simple, if it isn't, you are thinking about it too hard. It is simple now because we are using simple numbers. As algebra continues, these numbers will get more complicated. For example, let's use some terms instead of just whole numbers.

$$2xy^2 + 4xy^2 = 4xy^2 + 2xy^2$$

Or how about this for complicated:

$$(4a^2b^3 - 2a^2b^3) + (8a^2b^3 + 9a^2b^3) = (8a^2b^3 + 9a^2b^3) + (4a^2b^3 - 2a^2b^3)$$

So, even though it is simple now, later it will become a helpful tool. Just trust me, it will.

This law also applies when you have a bunch of numbers. For example, you can add 1 + 2 + 3 + 4 in any order and you will always get the same answer. Give it a try. I've rearranged the same four numbers four different ways. Add the following four problems.

$$1 + 2 + 3 + 4 = \qquad 2 + 4 + 1 + 3 =$$
$$4 + 1 + 3 + 2 = \qquad 3 + 1 + 2 + 4 =$$

The answer is always 10! That's what the *Commutative Law of Addition* is trying to teach you. The word "commute" means "to exchange," so basically it means you can exchange the numbers and still get the same answer. It's really that simple, if you know that 1 + 2 is the same thing as 2 + 1, then you understand the Commutative Law of Addition.

The next law is called the *Associative Law of Addition*. This one is just as simple. Here is the template for the Associative Law of Addition.

$$(a + b) + c = a + (b + c)$$

Basically, this law states that no matter how you group the numbers together, before you add them, you will still get the same answer.

The word "associative" means "a member of a group." So, this law is saying, "b" can hang out (associate) with "a" inside the parentheses or it can hang out with "c," either way you will get the same answer. Let's try it out. I'll put some numbers in place of a, b, and c.

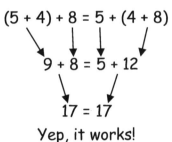

Yep, it works!

Let's review those two laws. The *Commutative Law of Addition* tells us that you can rearrange numbers in any order before adding them and you will always get the same answer. You can remember the word "Commutative" because the root word is *commute*. When you *commute* back and forth to work, you are "exchanging locations" the word *commute* means "to exchange." In this case, we are exchanging the location of the numbers.

The *Associative Law of Addition* tells us that you can group numbers together in any order before adding them and you will always get the same answer. To help you remember this law, look at the root word in "Associative." It is *associate*. To *associate* is to be "a member of a group." A business *associate* is someone you have paired up with to do business. An *association* is a group of people with a common interest. In this case, we are grouping up numbers into parentheses.

These laws may sound obvious or even useless, but later in algebra we are going to use these laws to help us solve big problems. This is as important as knowing the "ph" sound when learning to read. Just to make sure you fully understand these two laws, answer the following questions on the next worksheet.

Name_____ Date _____

Worksheet 17

1. $a + b = b + a$ is an example of which law?

2. $2a^2 + 5a^2 = 5a^2 + 2a^2$ is an example of which law?

3. What does the root word "commute" mean?

4. What does the root word "associate" mean?

5. $5 + (3 + 8) = (5 + 3) + 8$ is an example of which law?

6. $2x + (5x + b) = (2x + 5x) + b$ is an example of which law?

7. Use the Associative Law of Addition to rewrite the following expression.

    $(19 + a) + b =$

8. Use the Commutative Law of Addition, to rewrite the following.

    $7a + 2 + 3a =$

LESSON 18: LAWS OF MULTIPLICATION

Good News! The next two laws are so simple, I can explain them both in one sentence. The two laws of addition you just learned also work for multiplication. Those two new laws are called *The Commutative Law of Multiplication* and *The Associative Law of Multiplication*. I bet you already know what these laws mean just by reading their titles. Do these next two templates look familiar?

$$a \times b = b \times a$$

This is the template for the Commutative Law of Multiplication.

$$(a \times b) \times c = a \times (b \times c)$$

This is the template for the Associative Law of Multiplication.

The last two lessons in this book have been the long, long version of saying this: "When it comes to multiplication or addition, you can rearrange the numbers in any order, grouped together or not, and you will always get the same answer."

This is not the case with subtraction and division. Is 5 - 2 the same thing as 2 - 5? Is 10 ÷ 2 the same thing as 2 ÷ 10? No, they are not; these laws do not apply to subtraction or division.

Before we continue, you should be aware of the many different ways to write multiplication. You know that 3 x 5 = 15. Here are four different ways to write 3 x 5 = 15.

$$3 \cdot 5 = 15$$
$$3(5) = 15$$
$$3 * 5 = 15$$
$$(3)(5) = 15$$

When you throw a variable into the equation, there are even more ways. One of the reasons there are so many different ways is because it gets confusing when you use the letter "x" as a variable, next to a multiplication sign. Let's replace 3 from the last example with the letter b. Here are five different ways to write b x 5 = 5b.

$$5(b) = 5b$$
$$b \cdot 5 = 5b$$
$$(5)(b) = 5b$$
$$5b = 5b$$
$$b(5) = 5b$$

You must also be fully aware of the fact that a fraction is also a division problem! Did you know this? Look at the fraction below.

$$\frac{1}{4}$$

This fraction is read as one forth, BUT it is also read as 1 divided by 4. Do the math. I bet you'll come up with .25; because that's a quarter! This is true of every fraction; they are all division problems. With that in mind, look at this crazy fraction.

$$\frac{\frac{2}{9}}{\frac{3}{5}}$$

What is this fraction saying? Since a fraction is a division problem, I will rewrite it, so it looks more familiar.

$$\frac{2}{9} \div \frac{3}{5} = \frac{2}{9} \times \frac{5}{3} = \frac{10}{27}$$

See...a fraction on top of a fraction, in a fraction, isn't so crazy...it's kind of easy!

Try one by yourself.

$$\frac{\frac{1}{12}}{\frac{4}{13}}$$

I would suggest you rewrite it like I did. Don't forget the trick to dividing fractions; flip the second one and then multiply straight across. Do the math and then see if you got the same answer that I did. My answer is $\frac{13}{48}$.

Before we go any further, let's make sure you understand everything so far. You must understand all the questions on the next test, in order to continue.

Name_____ Date _____

CHAPTER 3 REVIEW TEST

Use the words Integer, Natural Numbers, or Neither to describe the following sets of real numbers.

1. a. .5 14 27 .75 0

   b. -2 5 25 -243 0 17

   c. 5 7 110 1 3 287

   d. -9 -136 -298 -600 ½

Simplify the following.

2. $2y^2z^3 + 5y^2z^3 =$

3. $9xyz \cdot 7x^2y^3z =$

4. $\dfrac{14d^3e^2f}{7def} =$

Solve the following problems.

5. $|-9| + |9| =$ 6. $|13| - |-5| =$

7. $|1/2| \times |-25| =$ 8. $-27 + -7 =$

9. $-5 \times 8 =$ 10. $17 - -5 =$

11. $-56 \div -7 =$ 12. $-45.9 - -13.8 =$

13. $-90 + 45 =$

Rewrite each expression using the Associative Law of Addition. The first one is done for you.

*. $(3a + a) + 5a = (3a + 5a) + a$

14. $7xy + (xy + 2) =$

15. $5 + (15 + 8) =$

Solve the following.

16. $\sqrt{\dfrac{16}{49}} =$

17. $\sqrt{\dfrac{b}{64}} =$

18. $\dfrac{\frac{5}{7}}{\frac{3}{4}} =$

19. $\dfrac{12a^4 - 6a}{3a^2} =$

Rewrite each expression using the Associative Law of Multiplication. The first one is done for you.

*. $6(2a \cdot 3) = 2a(6 \cdot 3)$

20. $(7b \cdot 3) \cdot 3b =$

21. $x \cdot (2x \cdot \tfrac{1}{2}) =$

CHAPTER 4   DISTRIBUTIVE LAW OF MULTIPLICATION

LESSON 19: MULTIPLYING A TERM BY AN EXPRESSION

The *Distributive Law of Multiplication* is used a lot in algebra. Sometimes, it is called the Distributive *Property* of Multiplication. You probably learned how to use this law in pre-algebra, but we will quickly review it because it is very important. I'll explain how this law works. Look at the problem below.

$$3(6 + 2) =$$

This problem can be solved two different ways. You could add the numbers in the parentheses (6 + 2) and then multiply that answer by 3.

$$3(6 + 2) =$$
$$3 \times 8 = 24$$

Or you could use the *Distributive Law of Multiplication* to solve this problem. This law states that you can multiply the outside number by the first number in the parentheses.

$$3(6 + 2) =$$
$$3 \times 6 = 18$$

Then multiply the outside number by the second number in the parentheses.

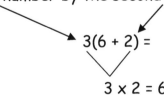

$$3 \times 2 = 6$$

Add those two answers together, 18 + 6 and again, you get 24. Do you see how you get the same answer whether you add the numbers first or you use the Distributive Law?

Using the Distributive Law of Multiplication may seem like more work, but sometimes it is your only choice. Sometimes you can't add the numbers in the parentheses. Look at this next problem and I'll explain what I mean.

$$5(4 + a)$$

We can't add (4 + a), so we have to use the Distributive Law. Multiply 5 by the first term and then by the second term. Now add those two answers together.

$$5(4 + a)$$
$$(5 \cdot 4) + (5 \cdot a) = 20 + 5a$$

Since 20 and 5a are not like terms, there is nothing more we can do here. Let's try another one together. Use the Distributive Law of Multiplication to simplify this next problem.

$$3a^2(3a + 4b) =$$

Remember, multiply the outside term by the first term in the parentheses. Then multiply the outside term by the second term. Since there is a plus sign, add the two answers together.

$$3a^2(3a + 4b) =$$

Here is how you should write it out on paper.
$3a^2(3a + 4b) =$
$(3a^2 \cdot 3a) + (3a^2 \cdot 4b) =$
$9a^3 + 12a^2b$

Try some on your own on the next worksheet.

Name_____ Date _____

Worksheet 19

Use the Distributive Property of Multiplication to solve the following.

1. $4(5 + 7) =$

2. $7(8 + a) =$

3. $8a(5a + 2) =$

4. $3x^2(2x + 9x^3) =$

5. $2xy(2xy + 5) =$

6. $2x^2y^2(3x + 4y) =$

7. $-7(5a + 8x) =$

8. $-10(a + b) =$

9. $6a^3b^2(-2ab + 3a^3b^2) =$

10. $-\frac{1}{2}a\left(-\frac{2}{3}a + \frac{3}{4}b\right) =$

11. $\frac{3}{8}b^2\left(-\frac{1}{4}b - \sqrt{\frac{4}{16}}\right) =$

LESSON 20: DISTRIBUTING A NEGATIVE NUMBER

Here is another situation where you will use distributing. The rules are the same, but it looks different. Look at the following expression.

$$-(5a - 4b)$$

Having a negative sign on the outside is the same thing as having an invisible -1 on the outside. Since there is a negative sign on the outside, you must multiply each term by -1.

$$-1(5a - 4b)$$

Here is the math. $(-1 \cdot 5a) - (-1 \cdot 4b)$
                      $-5a \quad - \quad -4b$

Fix the neg./pos. signs. $-5a + 4b$

Look back at the original problem and notice the difference between it and our answer. That negative sign that was on the outside has changed the sign of each term on the inside. The positive term is now negative and the negative term is now positive. So, the next time you see a negative sign in front of a set of parentheses, skip all the distributing and just change the sign of each term.

Here is another type of situation where that skill will come in handy. Sometimes in algebra you will have to subtract an expression from another expression. For example, look at this next problem.

$$(2xy + 5x - 9y) - (xy - 4x + 3y)$$

This problem is asking you to subtract $(xy - 4x + 3y)$ from $(2xy + 5x - 9y)$.

There are two ways to look at this problem. First, we will write it differently, so it is easier to understand.

$$\begin{array}{r} (2xy + 5x - 9y) \\ - (xy - 4x + 3y) \end{array}$$

Now we can read this more clearly. Reading vertically, in the direction of the arrows on the last page, you can see that this is actually 3 separate subtraction problems.

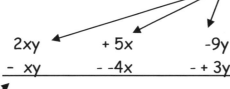

```
   2xy        + 5x        -9y
 - xy        - -4x       - + 3y
```

Look at this one first: 2xy – xy. Remember, "xy" is the same thing as 1xy. So, this problem is actually 2xy – 1xy, since these are *like terms*, we can just subtract. If 2 hats – 1 hat = 1 hat, then 2xy – 1xy = 1xy. Get it?

```
   2xy        + 5x        -9y
 - xy        - -4x       - + 3y
  ─────       ─────       ──────
   xy
```

Next we will tackle this problem: 5x - -4x

These are also like terms, so they can be combined too. The rule for subtracting a negative number is to turn both negative signs into positive signs. Our new problem is now 5x + +4x. Again, since these are like terms, just add them together 5x + 4x = 9x.

```
   2xy        + 5x        -9y
 - xy        - -4x       - + 3y
  ─────       ─────       ──────
   xy         + 9x
```

We are ready to solve the final subtraction problem, -9y – +3y

```
   2xy        + 5x        -9y
 - xy        - -4x       - +3y
  ─────       ─────       ──────
   xy         + 9x        -12y
```

The answer: xy + 9x – 12y

The second way to solve this problem is by using the shortcut we learned earlier. When there is a negative sign in front of a set of parentheses, you can just simply change the sign of each term in the parentheses and then try to solve it.

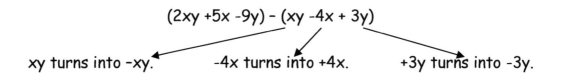

Now rewrite the problem vertically with the new signs and do the math.

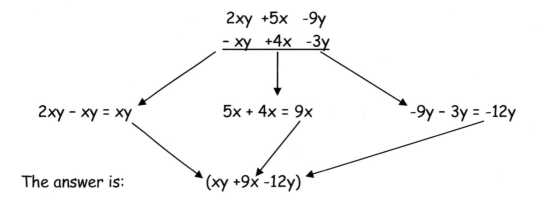

The answer is: (xy +9x −12y)

Just to make sure you fully understand this point, we are going to go over each step again.

→ -(xy −4x + 3y)

This negative sign is the same thing as −1 because 1 times anything is that number. So, let's rewrite that problem with a −1 in front.

$$-1(xy - 4x + 3y)$$

I will use the Distributive Law of Multiplication to multiply each term by −1, one at a time.

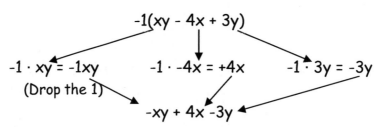

See! It's the same thing as just changing the signs.

Name_____ Date_____

Worksheet 20

Use the Distributive Property of Multiplication, to solve the following.

1. $-1(3y + 2x) =$

2. $-(5m - 7n) =$

3. $-(-7a - 9b) =$

4. $-1(3x + -5y) =$

5. $-2(a + b) =$

6. $-5a(5a - 2a) =$

7. $-x(x^2 - -y) =$

8. $3a(-a + -b) =$

9. $7(9 - 3) =$

10. $-10a(-3a + 7a^2) =$

11. $-\frac{3}{5}mn\left(-\frac{5}{6} + mn\right) =$

12. $-\sqrt[3]{64}(-\sqrt[3]{8} + -\sqrt[3]{27}) =$

LESSON 21: MULTIPLYING A TERM BY A TRINOMIAL EXPRESSION

That last worksheet had you multiply one term by one expression.

$$5a(2a^2 + 2x)$$

The expression in the example above is called a *binomial expression* because it contains two terms (bi means two). Next you will learn how to multiply a term by a *trinomial expression*: an expression with three terms (tri means three).

$$4a(2a + 5x + 3y)$$

The Distributive Law of Multiplication still applies here. Multiply the outer term by each of the terms inside the parentheses, one at a time. Let's get started.

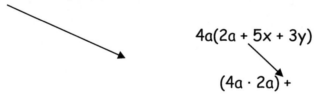

Start by multiplying this term by this term. For now, just write (4a · 2a) underneath the problem. We will solve it in a moment.

$$4a(2a + 5x + 3y)$$

$$(4a \cdot 2a) +$$

Next, bring down the plus sign.

Now multiply the outer term by the next term in the parentheses.

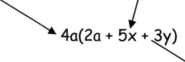

So far, your answer looks like this: (4a · 2a) + (4a · 5x) + Bring down the next sign and multiply the outer term by the last term in the parentheses.

$$4a(2a + 5x + 3y)$$

Here is how it looks after you distribute.

Do the math.

$$(4a \cdot 2a) + (4a \cdot 5x) + (4a \cdot 3y)$$
$$8a^2 + 20ax + 12ay$$

There are no like terms, so this problem is finished. Let's try another one together. This time there will be negative signs too. Try to solve it on your own, before looking at the solution.

$$5x(-3x + 4y - 5)$$

This one is a little more challenging because of the negative signs, but the rules are the same. Multiply the outer term by each inner term, one at a time. Look at the solution below.

Problem: $5x(-3x + 4y - 5)$
Distribute: $(5x \cdot -3x) + (5x \cdot 4y) - (5x \cdot 5)$
Answer: $-15x^2 + 20xy - 25x$

Since there are no "like terms," this problem is finished. Look at that last problem again and pay special attention to where I put the negative and positive signs.

$$5x(-3x + 4y - 5)$$
$$(5x \cdot -3x) + (5x \cdot 4y) - (5x \cdot 5)$$
$$-15x^2 + 20xy - 25x$$

If I had written $(5x \cdot -3x)(+5x \cdot 4y)(-5x \cdot 5)$ it would completely change the answer because when you put parentheses next to parentheses like that, it means $(-15x^2)$ **TIMES** $(20xy)$ **TIMES** $(-25x)$ and that is the wrong answer.

Be sure you are looking at the math sentence correctly. $5x(-3x + 4y - 5)$ is read as $5x$ times $(-3x + 4y - 5)$. When you apply the Distributive Law of Multiplication, it is read as $5x$ times $-3x$, *plus*, $5x$ times $4y$, *minus* $5x$ times 5. Make sure you fully understand this point. This is where a lot of people get lost in algebra, so if you can understand this concept, you will do well in algebra.

I know that last one was a little confusing because when I distributed the -3x, I put the negative sign inside the parentheses and when I distributed the -5, I didn't. Let's take one more look at the last example.

$$5x(-3x + 4y - 5)$$

$$(5x \cdot -3x) + (5x \cdot 4y) - (5x \cdot 5)$$

$$-15x^2 + 20xy - 25x$$

I could have put this negative sign on the outside, like I did with this negative sign. This is how that second line of math would have looked, had I put the negative sign on the outside instead.

$$-(5x \cdot 3x) + (5x \cdot 4y) - (5x \cdot 5)$$

Here is the math:

$$-15x^2 + 20xy - 25x$$

You get the same answer.

Let's try another one together.

$$7x(-3y + 8x - 9y)$$

Start by multiplying the term on the outside by each of the terms on the inside.

$$(7x \cdot -3y) + (7x \cdot 8x) - (7x \cdot 9y)$$

Do the math inside the parentheses.

$$(-21xy) + (56x^2) - (63xy)$$

Look for any like terms and combine them. I see two terms that are alike: -21xy and -63xy. I will group those two terms together.

$$(-21xy - 63xy) + 56x^2$$

Do the math inside the parentheses.

$$-84xy + 56x^2$$

Let's try one more. Try each step on your own, before reading the solution.

$$-4xy(3xy + 2x - 4y)$$

Multiply each term in the trinomial by the term in front. Since the coefficient of the term in front is a negative number, each term is multiplied by -4xy. Be sure to include that negative sign inside the parentheses.

$$(-4xy \cdot 3xy) + (-4xy \cdot 2x) - (-4xy \cdot 4y)$$

Do the math inside each set of parentheses.

$$(-12x^2y^2) + (-8x^2y) - (-16xy^2)$$

As you take away the parentheses, you will notice some extra negative and positive signs; you don't want to leave it like this. Use your negative/positive skills to fix it.

$$-12x^2y^2 + -8x^2y - -16xy^2$$

The first term won't change; it stays $-12x^2y^2$. The second term is "plus negative eight." That is the same thing as minus eight. The third term is "minus negative sixteen" the rule with a minus minus is to turn them into a plus plus. In this case, - - $16xy^2$ turns into ++ $16xy^2$, but drop one of the plus signs, there is no such thing as super positive.

$$-12x^2y^2 - 8x^2y + 16xy^2$$

We are going to do the exact same problem again, but this time we are going to solve it in our head and just write down the answer as we solve it. This way is much faster because we will include the negative and positive signs in our multiplication, eliminating the need for a bunch of parentheses. Let me explain.

$$-4xy(3xy + 2x - 4y)$$

To solve the problem above, what is the first thing you do? $-4xy \cdot 3xy$ right? In your head you should instantly see that the answer will be a negative number, so write that down. Next, you can solve 4 x 3 in your head, so write that down after the negative sign. All that is left is $xy \cdot xy$, which is x^2y^2. Write down all of that. So far, you should have written, $-12x^2y^2$ for your answer.

The next step is $-4xy \cdot +2x$. Do the math in your head. Start with -4 x 2, write down your answer. Multiply $xy \cdot x$ in your head, that's x^2y. So far, you should have written down $-12x^2y^2 - 8x^2y$.

The final step is $-4xy \cdot -4y$. When we solve the math in our head and just write down the answer, the signs are included in our multiplication, so we don't have to put them in between the sets of parentheses like we did before. By solving it in our head, we get to skip a step making it much faster.

Back to our problem, start with -4 x -4 and write down your answer. Multiply $xy \cdot y$ in your head, that's xy^2. That completes the problem, the answer is $-12x^2y^2 - 8x^2y + 16xy^2$.

Try to solve this next problem in your head. Write down your answer and then compare it to mine.

$$6x(-3y + 2x - 2a)$$

$6x(-3y + 2x - 2a)$
$6x(-3y) + 6x(2x) - 6x(2a)$
$-18xy + 12x^2 - 12ax$

That's some pretty cool math to do in your head. If you can do that, then you are ready for the next worksheet. If you are confused, you need to read the *Learn Math Fast System Volumes 1 - 4* first. It's time for you to try a few on your own.

Name_____ Date _____

Worksheet 21

Use the Distributive Law of Multiplication to solve the following problems.

1. $6x(-4y + 8x - 7y)$

2. $3a(4 - 6a + 2a^2)$

3. $-(-x + 4y - 3)$

4. $-6xy(5y + 2xy - 4y^2)$

5. $4(6 + 2x - 3x^2 - 6y)$

6. $-3xy^3(-4x^2y^2 + 2y - 6x)$

7. $7a(8a + b + 2c)$

8. $a^2(4x - a - 7x^3)$

Compare your answers to the ones in the back of the book. You must get 100% on this worksheet before moving forward. This is a very important concept in algebra. If you are confused now, you will want to go back because it's not going to get easier. Make sure you fully understand before continuing.

Name_____ Date _____

CHAPTER 4 REVIEW TEST

Simplify the following.

1. $5a(a + b)$

2. $2x^2(3x + 5)$

3. $-\frac{3}{8}a\left(\frac{3}{4}a + \frac{4}{7}b\right)$

4. $-(3a + 2a - 6b)$

5. $\begin{array}{r} 4y + 3a^2 + 10y \\ \underline{-2y - a^2 - 2y} \end{array}$

6. $3a(6b^2 - 4ab + 8)$

7. $5b(-3b - 7b + 3b^2)$

8. $-3x^2y^2(-8y^5 + 9x^5 - 6y^3)$

CHAPTER 5 FACTORING AND PRIME FACTORIZATION

LESSON 22: FACTORING

Earlier we talked about factors. In the problem 3 x 5 = 15, the 3 and 5 are the factors of 15. In the problem 6 x 4 = 24, the 6 and 4 are the factors of 24.

The number 24 has other factors too. For example, 8 and 3 are also factors of 24. Any two numbers or terms multiplied together are the factors of the product (the answer).

$$10 \times 6 = 60$$

Factor   Factor   Product

Now that you know what it takes for a number to be a factor, we will learn about *factoring*. Factoring is a math skill you use to find the factors of a particular number. For example, if I gave you the number 10, could you tell me two numbers that can be multiplied together to equal 10? The numbers 5 and 2 are factors of 10. That is factoring the number 10.

Can you think of any other factors of 10? You might be tempted to say 10 and 1, and you'd be right. But since 1 is a factor of every number, those two don't count. This is a pretty easy skill that you probably learned in elementary school. In algebra, it gets a little more challenging, so I'm starting off simple.

Let's find all the factors of the number 48. What two numbers equal 48 when multiplied together?

$$48$$
$$6 \times 8$$

Can you think of any other factors of 48? I know that 12 x 4 = 48, so those are also factors of 48. You could also multiply 24 x 2 to get 48. I will list all the possible factors of 48.

Factors of 48:   2, 4, 6, 8, 12, 24

Can you think of anymore factors of 48? I like to go through each natural number, staring at 2. I only have to go up to 24 because once I reach the halfway point, I know I have tried all possible factors.

As I go through each number, I stop at the number 3 and think…is 48 divisible by 3? I can tell you a quick trick to figure that out, besides dividing, of course. Add up the digits in the number 48, 4 + 8 = 12. If that answer is divisible by 3, then 48 is divisible by 3. Yep, 12 is divisible by 3, so 48 is too.

$$48 \div 3 = 16$$

Add both 3 and 16 to our list of factors.

Factors of 48:   2, 3, 4, 6, 8, 12, 16, 24

The numbers above are all the possible factors of 48. Now let's factor a number with a variable. What are the factors of $27x^2$?

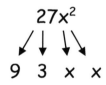

The factors of $27x^2$ are 9, 3, x, and x.

Try some on your own, by completing the next worksheet.

Name_____ Date _____

Worksheet 22

Find all the possible factors of the following numbers.

1. 30

2. 12

3. 55

4. 54

5. 24a

6. 9b

7. x^2

8. 18y

9. $7ab^2$

10. 64xyz

LESSON 23: PRIME NUMBERS

Another thing you probably learned in elementary school is prime factorization. We will review this quickly, to make sure you understand what prime numbers are and what prime factorization means.

There are a handful of numbers that don't have any factors besides 1 and itself. Those types of numbers are called *prime* numbers. Look at the numbers below.

$$4 \quad 5 \quad 6 \quad 7 \quad 8 \quad 9$$

The first number is 4. Can you tell me the factors of 4? They are 2 and 2 (and 1 and 4, but we don't care about those). Look at the second number, 5. What are the factors of 5? The only two numbers that can be multiplied together to equal 5 are 1 and 5. Since those are the only factors, 5 is a prime number.

The third number is 6. What are the factors of 6? They are 2 and 3, so 6 is not a prime number. How about the number 7? Is it a prime number or can you find factors of 7 besides 1 and 7? There are no two numbers you can multiply together to get 7, so it is a prime number. How about 8 or 9? Is either one a prime number? Try factoring 8 and 9 to find out. Did you come up with 4 x 2 and 3 x 3? That proves that 8 and 9 are not prime numbers.

If a number is not a prime number, it is a composite number. But that's not important to remember. I just thought you should know. It's funny, the number 1 is in a bit of a gray area. It's not prime, but it's not composite either. It's kind of in its own class. Let's not get hung up on that debate, it's not that important either.

The number 2 is also a weird one because it is the only even number that is prime. All other even numbers can be divided by 2, so they will always have factors besides 1 and itself. The number 2 is considered to be the FIRST and the SMALLEST prime number, and the only even number to be prime. Just a little trivia for you.

Name_____ Date _____

Worksheet 23

List all the prime numbers between 2-50.

Factor the following terms.

1. $3a$

2. $2y^2$

3. $11(a + b)$

4. $3ab^3c$

5. $7(8 - n)$

6. $5xyz^2$

7. $a(b)$

8. $41a^3b^3c$

9. pqr

10. 10^2

11. $\frac{1}{2} bh$

12. bhw

13. $d\pi$

14. πr^2

LESSON 24: PRIME FACTORIZATION

Now that you understand what it takes for a number to be prime; let's review prime factorization. I have listed three different sets of factors for the number 24.

Prime factorization means to keep factoring until all the factors are prime numbers. So, I will continue to factor each factor until we are down to prime numbers only.

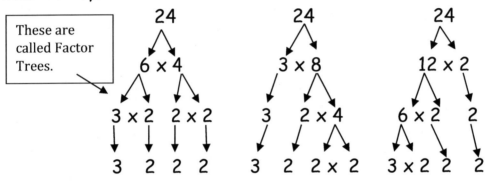

These are called Factor Trees.

Look at that! The prime factors of 24 are always 3, 2, 2, and 2. That's kind of cool, let's try another one. We will factor the number 36, three different ways, down to only prime numbers.

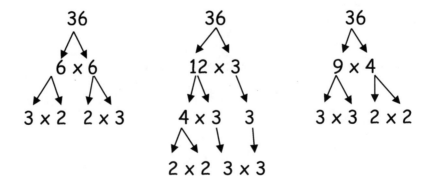

Look at that! No matter how you factor 36, you will always end up with the prime numbers 3, 3, 2, and 2. Try some on your own.

Name_____ Date _____

Worksheet 24

Factor the following numbers down to all prime numbers.

1. 18 2. 36 3. 49

4. 100 5. 62 6. 63

7. 77 8. 32 9. 47

10. 82 11. 52 12. 81

13. 48 14. 64 15. 65

LESSON 25: FACTORING OUT A COMMON FACTOR

That's enough of the elementary school review, let's get back to algebra. Earlier you learned how to factor a term. For example, what are the factors of this term?

$$3x^2y$$

The factors of $3x^2y$ are 3, x, x, and y. Try this one. What are the factors?

$$4(5 + a)$$

What is being multiplied? The number 4 is being multiplied by (5 + a). Those are the factors, 4 and (5 + a).

Knowing how to factor terms will make it easy for you to learn how to *factor out a common factor*. We will start off simple. So simple that it may seem pointless. But you will use this skill later to solve algebra problems, so you need to learn it.

Use the Distributive Law of Multiplication to simplify this term.

$$2(x + 2)$$

I've done the math for you.

$$2(x + 2)$$
$$2(x) + 2(2)$$
$$2x + 4$$

Now let's REVERSE that process. This time we will start with 2x + 4 and then we'll factor out the common factor, until we get back to 2(x + 2). What are the factors of 2x and 4?

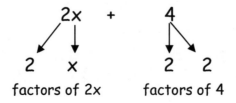

114

To "factor out" a number is to find a common factor of each term. That common factor is put on the outside of the parentheses. Look at the factor trees on the last page. Do you see a common factor between 2x and 4? I do, they both have 2 as a factor. Let's take that 2 and put it on the outside of a set of parentheses.

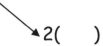

2( )

Look at those factor trees again. What factors are left over after taking a 2 from each term? The 2x is left with an x, and the 4 is left with a 2. Put those factors, along with the original sign, inside the parentheses.

2(x + 2)

And tada, we are back to our original problem. What we just did was nothing more than multiply 2(x + 2) and then we did the reverse math. Finding a common factor with a factor tree is actually a form of division. And division is the opposite of multiplication.

Let's try another one. First, we will multiply and then we will undo our multiplication by dividing. I've already distributed the terms for you.

3(3x + 8)
3(3x) + 3(8)
9x + 24

Now we will do the opposite operation. We will divide 9x + 24 by a common factor, so we can put it outside of the parentheses. Both 9x and 24 can be divided by 3.

9x ÷ 3 = 3x
24 ÷ 3 = 8

Put the 3, the common factor, on the outside of the parentheses and put the answers on the inside of the parentheses.

3(3x + 8)

Do you see how division or finding a common factor will get us back to our original problem? Did you notice that 3(3x + 8) is equal to 9x + 24? One is just turned inside-out from the other one. Here is another one.

$$5x(4x + 5)$$

Use the Distributive Law of Multiplication to simplify this problem.

$$5x(4x + 5)$$
$$5x(4x) + 5x(5)$$
$$20x^2 + 25x$$

Now we will do the opposite operation. Start by factoring each term down to PRIME factors. This is an easy way to find a common factor.

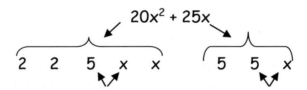

How many common factors do you see? I see that they both have 5 as a factor and they both have x as a factor. That will be our common factor, 5x. The next step is to divide each term by 5x, or just simply look at the factors that are left over and put those in the parentheses.

$$20x^2 \div 5x = 4x$$
$$25x \div 5x = 5$$

There you go, put the 5x on the outside of the parentheses and put the rest inside.

$$5x(4x + 5)$$

Let's try this a different way. Can you find the common factor in this example?

$$2xy + xy$$

Which factors do they have in common? They both have "xy" as a factor. Remove the xy from each term and put it on the outside of the parentheses.

> Remember: xy is the same thing as 1xy.

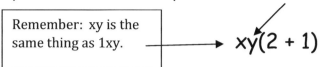

$xy(2 + 1)$

Try another one. What common factors do these two terms have?

$$x^2 + xy$$

They both have an x as a factor. Take one x out of each term. What are you left with in the parentheses?

$$x(x + y)$$

To make sure I am right, I will use the Distributive Law of Multiplication, to get back to the original expression.

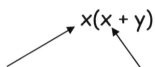

$x(x + y)$

Let's see...multiply the outside term by the first term, plus the outside term times the second term and I get $x^2 + xy$...yep! I got it right. The next example is even more challenging. Find the common factors.

$$2xy + 4xy$$

You may be tempted to say xy is the common factor, but don't forget about the coefficients, they are factors too. I will separate these two terms into their PRIME factors.

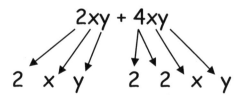

2xy + 4xy

2 x y   2 2 x y

Now it is easy to see what factors they have in common, 2xy. Let's put the common factors on the outside of the parentheses and put whatever is left over inside.

117

$$2xy(\quad)$$

Do you know what to put in the parentheses? The only number you can multiply 2xy by, to equal 2xy, is 1. That is the first term inside the parentheses.

$$2xy(1\quad)$$

Do you know what to do next? Should there be a plus sign or minus sign next? Look at the original problem, 2xy + 4xy. Since the original problem has a plus sign, use a plus sign.

$$2xy(1 + \quad)$$

What will the second term be? Look at the factor tree for anymore factors. There is still a 2 in the factor tree. That is the second term in the parentheses.

$$2xy(1 + 2)$$

Let's take a second look at the last original problem and the new version that has 2xy factored out.

$$2xy + 4xy \qquad 2xy(1 + 2)$$

If we've done our math correctly, these two should be equal. I'll do the math to find out.

$$2xy + 4xy = \mathbf{6xy} \qquad 2xy(1 + 2) =$$
$$2xy(3) = \mathbf{6xy}$$

Yep! They both equal 6xy, we have factored it correctly. You're probably asking yourself, "Why would we want to change a simple math problem into a more complicated one?" Because factoring is a skill you must learn, so you are able to solve complex algebra problems later.

Have you ever watched The Karate Kid movie? These lessons are like Danielson's "wax-on, wax-off" lessons. Eventually they will lead to something cool.

Name_____ Date_____

Worksheet 25

Factor out the common factors. The first one is done for you.

* $6a + 2a$ $2a(3 + 1)$

1. $5a + 10a$

2. $7x + 21x$

3. $x^2 + x^3$

4. $3xy - 6x$

5. $4ab - 2b$

6. $12xy^2 - 4y$

7. $3abc^3 - 9ab$

Next, look at each one of your answers from problems 1 – 7 and distribute each one below. The example from above has been done for you.

* $2a(3 + 1)$
     $(2a \times 3) + (2a \times 1)$
     $6a + 2a$

8.   12.

9.   13.

10.   14.

11.

Name_____ Date_____

CHAPTER 5 REVIEW TEST

Name each set of numbers as Integers, Natural Numbers, or Neither.

1.
   a. 555, 678, -945, -122
   b. 1, 2, 0, 4, 22, 11
   c. 956, 33, 25, 1,094
   d. $\frac{1}{4}$ $\frac{1}{2}$ $\frac{3}{4}$ $\frac{1}{1}$

Simplify the following.

2. $17xyz^3 - 12xyz^3 =$
3. $a^3b^2 + 2a^2b^2 =$
4. $6m^4n^3 \cdot 8mn^2 =$

Give the absolute value for each number.

5. $|-8| =$
6. $|23| =$
7. $|-\pi| =$
8. $|-d| =$

Do the math.

9. $-7 \cdot -8 =$
10. $-23 + -7 =$
11. $34 - -8 =$
12. $-11 \cdot 16 =$
13. $-3/4 \cdot 5/8 =$
14. $-m \cdot -n =$
15. $2a + -5a =$
16. $-3a(7a) =$

Chapter 5 Review Test continued

The following describes which law of addition?

17. $a + b = b + a$
18. $3a + 4a = 4a + 3a$
19. $2 + (3 + 4a) = (2 + 3) + 4a$
20. $8x^3y^2 + 3a + 5x^3y^2 = 5x^3y^2 + 8x^3y^2 + 3a$

The following describes which law of multiplication?

21. $2a \cdot 3a^2 = 3a^2 \cdot 2a$
22. $3a(2a \cdot 4x) = 4x(3a \cdot 2a)$
23. $5 \times 7 \times 10 = 7 \times 10 \times 5$
24. $6x(3x - 2a) = (6x \cdot 3x) - (6x \cdot 2a)$

Solve the following problems using any method you'd like.

25. $2(3 + 5) =$
26. $2a(4a + 6a) =$
27. $2x(5x - 3x) =$
28. $4m(3m - 2m) =$
29. $-(3y + 8x - 7y) =$
30. $-8x^2y(3x - 8xy^3) =$

Factor the following terms down to all prime factors.

31. $32x^2y$
32. $25y^2$
33. 42
34. $72ab$
35. $112x^3$

Chapter 5 Review Test continued

Factor out the common factor

36. $3\pi r^3 + 27r^3$
37. $16n^2 - 4n^3$
38. $49a^2b + 49ab^2$
39. $-3x^2 + 27x^2$
40. $\sqrt{7}ab + \sqrt{7}xy$

Use the Distributive Law of Multiplication, to solve the following.

41. $4xy(5y + 3x + 2)$

42. $-2a(9ab - 5b + a^2)$

43. $-\frac{5}{9}\left(ab^2 - \frac{4}{3}a^3\right)$

Divide the following terms.

44. $\dfrac{3x^2 + 6xy - 9x^3}{x^2} =$

45. $\dfrac{-3abc}{-12ab^2c^3} =$

46. $\dfrac{14y + 21xyz - 56y^3z^2}{7y^3} =$

Check your answers with the ones under Chapter 5 Review Test in the back of the book. This test should have been easy for you.

CHAPTER 6 POLYNOMIALS

LESSON 26: MONOMIALS, BINOMIALS, AND TRINOMIALS

A *Polynomial* (polly-no-me-ul) is a big, fancy word used to describe a group of terms. For example, this is a polynomial. This one has two terms:

$$(2y + 3)$$

When a polynomial contains only one term, it is called a *monomial*. Did you know that "mono" means 1. "Bi" means 2, and "tri" means 3? Just like **mono**tone means "one tone," a **bi**cycle has "two wheels," and a **tri**cycle has "three wheels." Can you guess how many terms are in a binomial? It is a polynomial with two terms. And of course, a trinomial has 3 terms. Here is an example of a trinomial expression.

$$2y - 2x + 6$$

A Trinomial is a Polynomial with 3 terms

Polynomials are a big part of algebra, so we will go over these thoroughly. But first, let's make sure you are up to speed on everything we've gone over, so far.

You already learned how to multiply a single term by an expression.

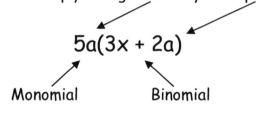

In other words, you learned how to multiply a monomial, by a binomial. And you already learned how to multiply a monomial, by a trinomial, in other words, one term by three terms.

$$(2x)(3x - 5 + 4x)$$

Monomial   Trinomial

123

But we will review it one more time, just to make sure you completely understand. You can probably guess how to solve this problem. You need to multiply each term in the trinomial by 2x. You should be able to solve this problem in your head. Start by multiplying the monomial by the first term in the trinomial.

$$2x(3x - 5 + 4x)$$

Can you solve 2x · 3x in your head? I can, it's $6x^2$. Write that down as the first part of the answer. Next, you need to multiply the second term in the trinomial, by 2x.

$$2x(3x - 5 + 4x)$$

Solve that in your head, 2x · -5 that equals -10x. Write down -10x after $6x^2$.

Now multiply the monomial, by the third term in the trinomial.

$$2x(3x - 5 + 4x)$$

You can solve that in your head too, 2x · 4x = $8x^2$. That is the last part of the answer. The last step is to look for any like terms.

$$6x^2 - 10x + 8x^2$$

Like terms

Combine the like terms and leave the rest alone. $6x^2 + 8x^2 = 14x^2$

Final answer: $14x^2 - 10x$

If that doesn't make sense to you, go back to the beginning or read Volume II and III of the *Learn Math Fast System*. If you are lost now, you will always be lost, but if you start over, you can catch up quickly.

Let's try another one. Again, we have a monomial times a trinomial.

$$7m(6m^2 + 8x - 6)$$

Try to solve this one in your head. What is the first step? 7m times what? That's right, $7m \cdot 6m^2$. That's easy to solve in your head. You know it is a positive number, so there is nothing to write yet. You know that 7 x 6 = 42, so that is the first thing to write down. Look at the m's. How many are there? There are 3 m's, so the first part of the answer is $42m^3$.

What is next? Multiply the monomial by the second term in the trinomial. That one is easy too. What is 7 x 8? You know it is 56. Just slap "mx" next to that and so far, the answer is $42m^3 + 56mx$.

Now multiply the monomial, by the last term in the trinomial. This one is simple too. What is 7 x -6? Slap an "m" on it, and the full answer is...$42m^3 + 56mx - 42m$. Are there any like terms? No, none of these terms are alike, so we are finished. Remember that m^3 and m are not like terms, so we cannot add them together. They can be multiplied together, but that is not the situation here.

Keep in mind, as you are trying to solve these problems in your head that you can look at them two different ways. I will use the same example I used above. You can solve that problem like this:

$$(7m \cdot 6m^2) \textbf{ PLUS } (7m \cdot 8x) \textbf{ MINUS } (7m \cdot 6).$$

$$42m^3 + 56mx - 42m$$

Or you can look at that same problem like this:

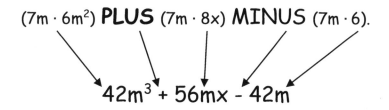

125

Either way you get the same answer. Just make sure you aren't (hooking)(parentheses) together without including the plus or minus signs in between because $(+42m^3)(+56mx)(-42m)$ is different than $(42m^3) + (56mx) - (42m)$.

Now it is your turn to multiply polynomials.

Name_____ Date _____

Worksheet 26

1. $4r(6r - 9x + 5t)$

2. $7x(8y + 3x - 2x^2)$

3. $-5y(3y + 2x - 4y)$

4. $-(9x + 10x^2 - 5y)$

5. $3x^2y(2x + 4y - 1)$

6. $2s^2(3s^2 - 5s + 6)$

7. $5a^2(5a - 3a^2 + 4)$

8. $-4a(a^2 - 2ab - b)$

9. $(-3xy)(x^2 - 2xy + y^2)$

10. $(6m^2 - 2n^2 - 4mn)2m^3$

LESSON 27: MULTIPLYING BINOMINALS

Next, we will learn how to multiply a binomial by another binomial. Look at this next problem.

$$(2x + 5)(x + 3)$$

Binomial   Binomial

The trick here is to ignore the second term in the first binomial for a moment.

$$(2x + 5)(x + 3)$$

Now that we are ignoring the "+5," we can multiply the same way we did before.

$$(2x + 5)(x + 3)$$

This times this

Do the math: $2x \cdot x = \mathbf{2x^2}$.

$$(2x + 5)(x + 3)$$

This times this

Now multiply the next two. Do the math: $2x \cdot 3 = \mathbf{6x}$. Here is your answer, so far: $2x^2 + 6x$...

Now we will ignore the 2x, bring back the +5, and repeat the process.

$$(2x + 5)(x + 3)$$

This times this.

Do the math: $5 \cdot x = 5x$

$$(2x + 5)(x + 3)$$

And then this times this.

Do the math: 5 × 3 = 15.

Put those four answers together, 2x² + 6x + 5x + 15, and look for any like terms. I see some like terms, +6x and +5x are alike. Put those two together. The final answer is:

$$2x^2 + 11x + 15$$

This is how you should write that problem on paper.

(2x + 5)(x + 3)
(2x · x) + (2x · 3) + (5 · x) + (5 · 3)
2x² + 6x + 5x + 15
2x² + 11x + 15

Let's try another one together. This time we'll skip this step by doing some of the math in our mind.

$$(5x^2 - 8)(3xy - \tfrac{1}{2})$$

Step 1:   Start with this   times this.

Both of those terms are positive numbers, so the first thing to solve is $5x^2 \times 3xy =$. That answer will be the first chunk of the whole answer.

(5x² - 8)(3xy - ½)

5x² · 3xy = 15x³y

Answer, so far is 15x³y

Step 2:

$$(5x^2 - 8)(3xy - \tfrac{1}{2})$$

Multiply this by this.

$$5x^2 \times -\frac{1}{2} = -\frac{5}{2}x^2$$

$$\boxed{\frac{5}{1} \times -\frac{1}{2} = -\frac{5}{2}}$$

Let's add that answer to our previous answer:

$$15x^3y - \frac{5}{2}x^2$$

Step 3:

$$(5x^2 - 8)(3xy - \tfrac{1}{2})$$

Multiply this by this.

$$-8 \times 3xy = -24xy$$

Add the answer to this problem to our big answer and we get $15x^3y - \frac{5}{2}x^2 - 24xy$.

Step 4:

$$(5x^2 - 8)(3xy - \tfrac{1}{2})$$

Multiply this by this.

$$-8 \times -\frac{1}{2} = \quad -\frac{8}{1} \times -\frac{1}{2} = \frac{8}{2} = 4$$

Add the above answer to our big answer to finish the problem.

$$15x^3y - \frac{5}{2}x^2 - 24xy + 4$$

Any like terms? Mmm…no, we are finished.

Since that math gets a little confusing, a lot of math people use the word FOIL to help them multiply binomials; let me explain. The word F.O.I.L. stands for First, Outer, Inner, Last. That's the order of multiplication; I'll explain what I mean.

FIRST: $(2x + 2)(x - 4) =$

Multiply the FIRST terms in each binomial, $2x \cdot x = 2x^2$. Write down that answer.

OUTER: $(2x + 2)(x - 4) =$

Multiply the two OUTER terms $2x \cdot -4 = -8x$

INNER: $(2x + 2)(x - 4) =$

Multiply the two INNER terms $2 \cdot x = 2x$

LAST: $(2x + 2)(x - 4) =$

Multiply the LAST terms in the parentheses together, $2 \times -4 = -8$

Put all those answers together. If there is no sign in front of the answer, it is assumed to be positive, so put a plus sign in front of it.

$$2x^2 \; -8x \; +2x \; -8$$

Combine any like terms, -8x and 2x are like terms. Together they are -6x, so now the final answer is $2x^2 -6x -8$.

Let's use the FOIL method, to solve this next problem.

$$(3y + 2)(y - 3) =$$

FIRST: $3y \cdot y = 3y^2$

OUTER: $3y \cdot -3 = -9y$

INNER: $2 \cdot y = 2y$

LAST: $2 \cdot -3 = -6$

Put all those answers together with their signs and then combine any like terms.

$$3y^2 - 9y + 2y - 6$$
$$3y^2 - 7y - 6$$

Let's try another one, using the FOIL method. Only this time, try to solve it in your head and just write down the answers.

$$(5m + 4)(m + 3)$$

Think to yourself, "Multiply the FIRST terms, $5m \cdot m$," and write down your answer.

Now multiply the OUTER terms, $5m \cdot 3$ and write down that answer.

Now multiply the INNER terms, $4 \cdot m$ and write down that answer too.

Finally multiply the two LAST terms, 4×3. Is this what your answer looks like?

$$5m^2 + 15m + 4m + 12$$

See any like terms? I do, add $15m + 4m$ and write the final answer, $5m^2 + 19m + 12$.

The FOIL method is quicker because instead of ending up with something like this: $(15x^3y) - (2.5x^2) - (24xy) - (-4)$, you end up with $15x^3y - 2.5x^2 - 24xy + 4$, right from the start. That's why you have to include the negative signs (if there are any) in your multiplication, when you are using the FOIL method.

Let me explain with this next problem. I will do the math using both methods, so you can see the steps that you skip with FOIL.

$$(5x - 2)(3x - 4)$$

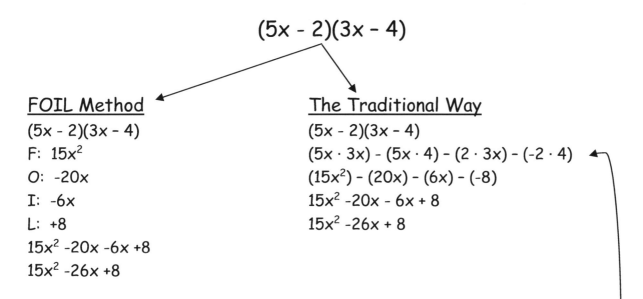

FOIL Method
$(5x - 2)(3x - 4)$
F: $15x^2$
O: $-20x$
I: $-6x$
L: $+8$
$15x^2 - 20x - 6x + 8$
$15x^2 - 26x + 8$

The Traditional Way
$(5x - 2)(3x - 4)$
$(5x \cdot 3x) - (5x \cdot 4) - (2 \cdot 3x) - (-2 \cdot 4)$
$(15x^2) - (20x) - (6x) - (-8)$
$15x^2 - 20x - 6x + 8$
$15x^2 - 26x + 8$

I don't know about you, but I prefer the FOIL method. It's much easier to keep track of the negative and positive answers, especially when you get math like this.

Take a look at this problem.

$$(x - 5)^2$$

Can you figure out what this means?

Think about what it means to square a "number." It means, "That number times itself." When you look at $(x - 5)^2$ it means $(x - 5)$ times $(x - 5)$. I'll write it out that way.

$$(x - 5)(x - 5)$$

That looks like a binomial times a binomial to me, so let's use the FOIL method to figure it out.

First: $x \cdot x = x^2$
Outer: $x \cdot -5 = -5x$ (Notice I brought the negative sign with the 5.)
Inner: $-5 \cdot x = -5x$
Last: $-5 \cdot -5 = 25$

Put the answers together.
$$x^2 - 5x - 5x + 25$$
Combine the like terms.
$$x^2 - 10x + 25$$

And that is the answer to $(x-5)^2$.

But there is a faster way. Mathematicians noticed that every time you solve a problem like $(a+b)^2$, the answers are always similar. Take a look at all of the answers below.

> Note: ab and ba are the same, but mathematicians insist we write them in alphabetical order.

$(a + b)^2 = (a + b)(a + b)$ = $a^2 + ab + ba + b^2$ = $a^2 + 2ab + b^2$
$(x - 5)^2 = (x - 5)(x - 5)$ = $x^2 - 5x - 5x + 5^2$ = $x^2 - 10x + 25$
$(4 - a)^2 = (4 - a)(4 - a)$ = $4^2 - 4a - 4a + a^2$ = $16 - 8a + a^2$

Do you see how all the answers look similar? They all start with the first term squared and end with the second term, from the parentheses, squared. The middle term, in the answer, is always the two terms multiplied together and then times two. With that in mind, try to solve this one in your head without using the foil method.

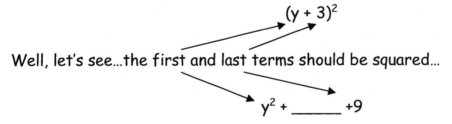

And what goes in the center? The two terms multiplied and doubled. So, let's see...3 times y times 2. That's 6y! Is this the answer you got?

$$y^2 + 6y + 9$$

Mathematicians write out this shortcut like this:

$$(a + b)^2 = a^2 + 2ab + b^2$$

If you can remember that shortcut, you can save some time. There are a lot of these shortcut formulas for things that come up often; they are called *identities*. One of the most famous identities is called the quadratic formula, which you will learn about later in this book.

Practice doing this by completing the next worksheet.

Name_____ Date _____

Worksheet 27 The F.O.I.L. Method

Use the FOIL method to simplify the following.

1. $(3x - 2)(4x - 1)$

2. $(5x - 3)^2$

3. $(x + y)(x - y)$

4. $(5y + 2)(5y - 2)$

5. $(a + b)(a - b)$

6. $(2a + 7)(a - 3)$

7. $(3x - 2y)^2$

8. $(3m - 1)(2m - 5)$

9. $(2a + b)^2$

10. $(m + n)^2$

LESSON 28: MULTIPLYING WITH TRINOMIALS

Compare your answers with the ones in the back of the book. If you did well, you are ready to continue. If you got one wrong answer, find out why. If you got several wrong answers, you should go back to where it was easy and start over.

Next, we will multiply a binomial by a trinomial, this should be easy too.

$$(3x - y)(2x^2 + 6x - y)$$

First, we multiply $3x$ by each of the terms in the trinomial, one at a time. This time I will use parentheses to show the multiplication of each term.

$$(3x - y)(2x^2 + 6x - y)$$
$$3x(2x^2) + 3x(6x) - 3x(y)$$
$$6x^3 + 18x^2 - 3xy$$

Now multiply "$-y$" times each term in the trinomial and tack that on to the end of our last answer.

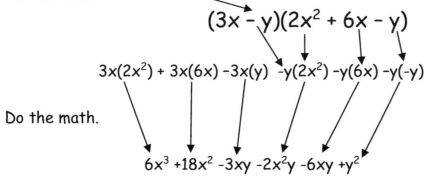

Do the math.

$$6x^3 + 18x^2 - 3xy - 2x^2y - 6xy + y^2$$

Let's rearrange these real quick, so it's easier to find any like terms.

I will put them in order from the highest number exponent to the lowest.

$$\underbrace{6x^3}_{3} + \underbrace{18x^2 - 2x^2y}_{2's} + \underbrace{y^2 - 6xy - 3xy}_{1's}$$

Now it is easy to see our like terms, let's combine them.

The final answer is: $6x^3 + 18x^2 - 2x^2y + y^2 - 9xy$

Next, you will multiply a trinomial by a trinomial.

$$(x^2 - 5x + 2)(3x^2 - x - 4)$$

The first step is to multiply x^2 by each term in the second trinomial, one at a time. Then multiply -5x by each term in the trinomial. And finally multiply each term by 2.

Problem: $(x^2 - 5x + 2)(3x^2 - x - 4)$
Distributed: $3x^4 - x^3 - 4x^2 - 15x^3 + 5x^2 + 20x + 6x^2 - 2x - 8$
Rearranged for clarity: $3x^4 - x^3 - 15x^3 - 4x^2 + 5x^2 + 6x^2 + 20x - 2x - 8$

Combine the like terms: $3x^4 - 16x^3 + 7x^2 + 18x - 8$

If you can handle that, you are ready for the next lesson: a binomial times a binomial times another binomial.

$$(2x + 3)(x - y)(4x + 2)$$

The trick here is to ignore the last binomial for a moment and use the FOIL method to simplify the first two binomials.

(2x + 3)(x - y) (4x +2)

First	Outer	Inner	Last
$2x^2$	$-2xy$	$+3x$	$-3y$

Squish those answers together and bring back the binomial we were ignoring.

$$(2x^2 - 2xy + 3x - 3y)(4x + 2)$$

If you prefer, you can switch those two sets of parentheses, so the (4x + 2) is in front. I know I can do this because the Commutative Law of Multiplication says it's cool to do that. I prefer to have the binomial in front, so I'm going to switch them around.

$$(4x + 2)(2x^2 - 2xy + 3x - 3y)$$

Now it is clear to see that we need to multiply 4x by each term and then multiply +2 by each term. Let's get started.

Problem: $(4x + 2)(2x^2 - 2xy + 3x - 3y)$

Distribute: $4x(2x^2) - 4x(2xy) + 4x(3x) - 4x(3y) + 2(2x^2) - 2(2xy) + 2(3x) - 2(3y)$

Do the math: $8x^3 \quad -8x^2y \quad +12x^2 \quad -12xy \quad +4x^2 \quad -4xy \quad +6x \quad -6y$

Rearrange and combine like terms: $8x^3 - 8x^2y + 16x^2 - 16xy + 6x - 6y$

Isn't that some unbelievable math to solve in your head? It looks really complicated, but it's actually pretty easy math. I think the hardest part is keeping the negative and positive signs straight. I notice that when I solve the multiplication in my head, I can just write down the answer with the correct sign included. I think that way is much easier than distributing the terms and writing everything out with parentheses.

Look at the problem above again. Try to solve the math in your head and come up with the answers that are on the "Do the math" line, skip writing out the distribution. That way makes it less confusing for me. You must *understand* the "Distribute" line of math, but you don't have to write it down, unless it helps you.

You've probably noticed by now that there are a lot of different ways to write the same thing in algebra. It's a lot like a computer. To get to a file on a computer, you can use a shortcut icon on your desktop, or you can go straight to the program and then open the file. Either way will get you the same file, but it is up to you how you get there. Algebra is the same way, you can write 3 x 2 or 3(2) or 3 · 2 or 2(2 + 1), and the answer will always be the same.

If all these negative and positive multiplication problems are mindboggling to you, you should read Volume II of the *Learn Math Fast System*.

Name_____ Date _____

Worksheet 28

Use the Distributive Law of Multiplication to simplify the following.

1. $3a(5a + 6x - 7)$

2. $(2 + 3x)(2x - 8y)$

3. $-7m(m + n + 2x)$

4. $(4x + 5)(3x - 1)$

5. $(3xy^2 + 4y)(4x + 3xy - 2x)$

6. $(x - 5)^2$

7. $(7y + 3)^3$

8. $(5xyz + 3z - 8y)(6z - 9x + 5xyz)$

9. $-(7x^2y^3z^4 - 12xy^3 - 5x^2y^3z^4)$

10. $(x - y^2)(y - x^2)(xy + xy)$

LESSON 29: FACTORING A QUADRATIC

Look at the equation below.

$$(5x + 3)(2x + 3) = (10x^2 + 15x + 6x + 9)$$

Factor   Factor   Product

The binomials $(5x + 3)$ and $(2x + 3)$ are the factors of $10x^2 + 15x + 6x + 9$. What are the factors of this problem?

$$(7x + 2)(2x + 4)$$

The factors are the "numbers" being multiplied. In this case, $(7x + 2)$ is being multiplied by $(2x + 4)$, so those are the factors. I will use the FOIL method to multiply these two factors together. Try to come up with the answer yourself, before you look at my answer.

$$(7x + 2)(2x + 4) = 14x^2 + 28x + 4x + 8$$
Combine the like terms.
$$14x^2 + 32x + 8$$

That is the product of those two factors. Guess what? When you multiply a binomial expression by another binomial expression, the answer is a *quadratic* expression. Get it? Quad means 4, and bi means 2, so 2 x 2 = 4, and a binomial times a binomial = a quadratic. Whenever you multiply two binomials, the answer will always fit into this mold. Do you see how the answer above looks a lot like this quadratic form?

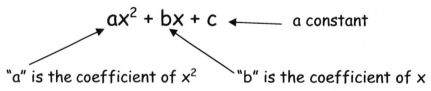

$ax^2 + bx + c$ ← a constant

"a" is the coefficient of x^2 "b" is the coefficient of x

The letter "a" is the coefficient of x^2. Sometimes the coefficient will be the number 1, which doesn't need to be written, so "a" may be invisible at times.

$$ax^2 + bx + c$$

The letter "b" represents the coefficient of "x" or whatever variable is being used. And the letter "c" is any number by itself, no variable; it's called a constant.

Whenever you have a polynomial that fits into this exact format, it is called a quadratic. And remember, a **QUAD**ratic is the product of two **BI**nomials.

Now, it is time for the next big challenge in algebra. It is called *factoring a quadratic expression.*

To factor a quadratic means to do the reverse of what we just did. We just took two binomials (7y + 2) and (2y + 4), multiplied them together, and ended up with the quadratic, $14y^2 + 32y + 8$. This time, we will start with the quadratic and then we have to figure out what two binomials were multiplied together to get that product.

Let's start with a simple one. I'll give you the quadratic and then we'll figure out what two binomials (factors) were multiplied together to get that product. We start with two empty sets of parentheses. We will fill in the blanks as we figure them out. Let's factor this quadratic expression.

Quadratic: $x^2 + 6x + 8$

Start with two empty sets of parentheses.   (   )(   )

Take a look at the first term, x^2. What two factors were multiplied together, to get x^2 as a product? The two factors of x^2 are x and x. So that must be the two "first" terms.

Quadratic: $x^2 + 6x + 8$

Put the two factors of x^2 in the parentheses.   (x   )(x   )

Now take a look at the last term in the quadratic, 8. When we use the FOIL method, the last step we do is multiply the LAST numbers together. This "8" is a product of the two LAST numbers in the parentheses. What two numbers do you suppose were multiplied together to get 8? Two factors of 8 are 4 and 2. Let's put those two numbers in the parentheses, and see if it makes sense.

Quadratic: $x^2 + 6x + 8$

Put the two factors of 8 in the parentheses.   (x   2)(x   4)

Since we have all positive number in our quadratic, it is safe to put "+" signs inside the parentheses too.

$$x^2 + 6x + 8$$
$$(x + 2)(x + 4)$$

We can find out if we are right, by using the FOIL method. Multiply those two binomials to see if the product (answer) is the quadratic we started with. Here is the math to check our work.

$$(x + 2)(x + 4)$$
$$x^2 + 4x + 2x + 8$$
Combine like terms.
$$x^2 + 6x + 8$$

Yep! When we multiply (x + 2) by (x + 4), we get a product of $x^2 + 6x + 8$. Now let's say we didn't use 4 and 2 for factors of 8. Let's say we chose 8 and 1, to be the factors of 8. Our binomials would have been (x + 8)(x + 1). Use the FOIL method to get the product of these two binomials instead.

$$(x + 8)(x + 1)$$
$$x^2 + 1x + 8x + 8$$
Combine like terms.
$$x^2 + 9x + 8$$

That answer is not the same as the quadratic we started with, so it is not the right answer. That's how I know that 4 and 2 are the right choice, not 8 and 1. Let's try another one. We will start with a quadratic and an empty set of parentheses.

$$x^2 + 8x + 15$$
$$(\quad)(\quad)$$

Look at the first term in the quadratic, x^2. What are the factors of x^2? Put those in the parentheses.

$$x^2 + 8x + 15$$
$$(x\quad)(x\quad)$$

Look at the last term in the quadratic, 15. What are the factors of 15? Let's see...5 and 3 are factors of 15, so are 1 and 15. Look at the term in the center of the quadratic, 8x. Which of those two sets of factors will total 8 when combined? Will 5 and 3, or 15 and 1? Definitely 5 and 3, put those in the parentheses.

$$x^2 + 8x + 15$$
$$(x\quad 5)(x\quad 3)$$

All the numbers are positive, so it is safe to put "+" signs in the parentheses.

$$x^2 + 8x + 15$$
$$(x + 5)(x + 3)$$

We can find out if we are right by multiplying the two binomials, to see if we come up with the same quadratic.

$$(x + 5)(x + 3)$$
$$x^2 + 3x + 5x + 15$$
$$x^2 + 8x + 15$$

Yep, we got the right answer, so our factors are correct. Let's try another one. This time, there will be a negative number involved.

Here is the quadratic expression.

$$x^2 - 7x + 12$$

Start with the first term in the quadratic, x^2. What two factors can produce x^2? It must be x and x again. Put those in the empty parentheses.

$$x^2 - 7x + 12$$
$$(x\quad)(x\quad)$$

The last term in the quadratic is 12. What are the possible factors of 12? I can think of two possible sets of factors, 6 and 2, and 3 and 4. Look at the center term and decide which set would work best. The center term is -7x, so I think 3 and 4 would be the best choice. Put those two numbers in the parentheses.

$$x^2 - 7x + 12$$
$$(x\quad 3)(x\quad 4)$$

But wait! The center term is *negative 7*, can you figure out what signs you will need to use for the 3 and 4?

Keep in mind that this amount is a *sum*, and this amount is a *product*.

$$x^2 - 7x + 12$$

A *sum* is the answer to an addition problem. A *product* is the answer to a multiplication problem. Adding negative numbers, gives a negative sum. Multiplying negative numbers gives a positive product.

In the problem above, we want the sum of 3 and 4 to be -7 and we want the product of 3 and 4, to be positive. That's easy, make them both negative!

$$-3 + -4 = -7$$
$$-3 \times -4 = +12$$

Fill in the parentheses and then use the FOIL method to see if we have the right factors.

$$(x - 3)(x - 4)$$
$$x^2 - 4x - 3x + 12$$
Combine like terms.
$$x^2 - 7x + 12$$

Yep, that's the right answer. Do you see how you can look at the factors of the last term and then decide which ones would be best by looking at the center term? Let's try another one. Here is the quadratic and an empty set of parentheses.

$$y^2 - 10y + 21$$
$$(\quad)(\quad)$$

What are the factors of y^2? Put those two factors in the parentheses.

$$y^2 - 10y + 21$$
$$(y\quad)(y\quad)$$

Look at the last term. What are the possible factors of 21? Keep in mind that the center term is -10y, so whatever factors you choose, they better equal -10 when combined. The possible factors of 21 are 7 and 3, or 1 and 21. Which two will equal 10? Definitely 7 and 3, just make them negative, so when they are multiplied together, they will be positive, and when they are added together, they will be negative.

$$y^2 - 10y + 21$$
$$(y - 7)(y - 3)$$

Use the FOIL method, to see if we got the right answer.

$$(y - 7)(y - 3)$$
$$y^2 - 3y - 7y + 21$$
Combine like terms
$$y^2 - 10y + 21$$

That's right! Let's try one more. This time our quadratic will have all negative numbers. If the last term, which is a product, is a negative number, then the two factors MUST have opposite signs.

Here is the next quadratic and an empty set of parentheses to start with.

$$m^2 - 3m - 18$$
$$(\quad)(\quad)$$

What are the factors of m^2?
What are the possible factors of 18?
Choose factors of 18 that total -3 when combined.
The possible factors of 18 are, 9 and 2, 6 and 3, or 18 and 1.
Which set of factors will total -3 when combined?
Remember the last term is negative, so you will have two different signs.
If we use -6 and +3, we will have a total of -3, so let's use those.

$$m^2 - 3m - 18$$
$$(m - 6)(m + 3)$$

Check our work with the FOIL method.

$$(m - 6)(m + 3)$$
$$m^2 + 3m - 6m - 18$$
$$m^2 - 3m - 18$$

That's it! Now try some on your own on the next worksheet. You will be given a quadratic and then you have to figure out what two binomials were multiplied together to get that quadratic. In other words, factor the quadratic.

Name_____ Date _____

Worksheet 29 Factoring Quadratics

Factor these quadratic expressions.

1. $x^2 + 8x + 16$

2. $m^2 + 5m + 6$

3. $x^2 - 8x + 7$

4. $y^2 + 4y - 32$

5. $x^2 + 13x + 42$

6. $n^2 + 5n - 36$

7. $b^2 + 7b - 30$

8. $c^2 - 10c + 16$

9. $a^2 + 8a - 33$

10. $x^2 + 16x + 48$

Name_____ Date _____

CHAPTER 6 REVIEW TEST

Solve the following.

1. $-x^2 \cdot -x^3 =$
2. $(-8) - (-5) =$
3. $4xy(-3xy) =$
4. $-3x^3(7x) =$
5. $-5(5x^4) =$
6. $-4x^2y^2 \cdot -4x^2y^2 =$
7. $-mn(m - n)$

Name the law that is being used to rearrange each problem.

8. $4a + 3 + 2a = 2a + 4a + 3$
9. $8a + 7 + 9 + 6a = (8a + 6a) + (7 + 9)$
10. $7xy(3x \cdot 2xy) = 3x(7xy \cdot 2xy)$
11. $(x)(5)(2) = 5 \cdot 2 \cdot x$
12. $4a(3x \cdot 9a) = (4a \cdot 3x)(4a \cdot 9a)$

Simplify the following using the Distributive Property of Multiplication.

13. $2a(5 - 3a)$
14. $-5b(2a - 3b)$
15. $m(2m + 4 + 6m^2)$
16. $-5x(3x + 4y - 4x + 3)$
17. $(b + c - d)a$

Chapter 6 Review Test page 2

Use the FOIL method to multiply the following.

18. $(b + 3)(b - 4)$
19. $(x - 4)(x - 5)$
20. $(y + 8)(y + 7)$
21. $(r - 9)(r + 2)$
22. $(m - 1)(m - 8)$

Factor the following polynomials.

23. $x^2 + 15x + 56$

24. $x^2 + 5x - 24$

25. $x^2 - 15x + 50$

26. $y^2 + 14y + 24$

27. $x^2 - 12x + 20$

28. $m^2 + 2m - 48$

29. $x^2 + 5x - 36$

30. $x^2 - 12x - 13$

CHAPTER 7 QUADRATIC EQUATIONS

LESSON 30: RECOGNIZING A QUADRATIC EQUATION

All of the problems from the last worksheet looked similar. Each one of them followed the same pattern. The coefficients and the variables may have been different, but basically, they all looked something like this quadratic form:

$$ax^2 + bx + c$$

On the last worksheet, the coefficient of "a" was always the number 1, which doesn't need to be written, so our "a" was invisible.

Once you add "equals something" to a quadratic polynomial, you have a *quadratic equation*. Here is the Standard Form of a Quadratic Equation:

$$ax^2 + bx + c = 0$$

Sometimes you will run across an equation that looks like this one. It looks similar to the standard form of a quadratic, but not exactly.

$$x^2 - 35 = 2x$$

But look a little closer; we can rearrange these terms, so they fit into our quadratic mold. Subtract the 2x from the right side and put it on the left side, in the center.

$$\begin{aligned} x^2 - 35 &= 2x \\ -2x &-2x \\ \hline x^2 - 2x - 35 &= 0 \end{aligned}$$

Now it looks like a quadratic equation. On the next worksheet, you will practice rearranging some equations, so they fit into our quadratic mold. You'll want your equation to match up to this one:

$$ax^2 + bx + c = 0$$

The first term may or may not have a coefficient, but it does need to have a squared variable.

$$ax^2 + bx + c = 0$$

The second term will have the same variable as the first term and it may or may not have a coefficient, or it may be zero.

$$ax^2 + bx + c = 0$$

The third term will be a number, by itself. This number is called a *constant* because it doesn't vary like a variable; this number is written in stone.

$$ax^2 + bx + c = 0$$

The equal sign with a number is what makes it an equation. On the next worksheet, you will be given equations that need to be rearranged, so they fit into our *Standard Quadratic Equation:*

$$ax^2 + bx + c = 0$$

Name_____ Date _____

Worksheet 30

Rearrange the equations, to fit into $ax^2 + bx + c = 0$. Write down which term is a, which is b, and which is c. The first one has been done for you.

* $-8x + 15 = -x^2$
         $x^2 - 8x + 15 = 0$ a = 1, b = -8, c = 15

1. $x^2 + 6x = -9$

2. $x^2 - 5 = 6x$

3. $8x + 12 = -x^2$

4. $m^2 - 4m = -4$

5. $y^2 - y = 20$

6. $a^2 - 30 = a$

7. $12x + 32 = x^2$

8. $z^2 - 28 = -3z$

LESSON 31: SOLVING FOR X IN A QUADRATIC EQUATION

Now that you can recognize a quadratic equation, let's learn how to solve for x in one. You will learn three different ways to do this. During this lesson, you will learn how to solve simple quadratic equations by hand. This may seem a little confusing at first, but you will quickly see that it is actually very easy math. Look at the example below. This is a quadratic equation.

$$x^2 + 4x + 4 = 4$$

This is the answer to our quadratic equation. We are trying to figure out what "x" equals to make this equation equal 4. The first step is to factor the quadratic. You know how to do that! I will put some empty parentheses below the equation for us to fill in with our factors.

$$x^2 + 4x + 4 = 4$$

$$(\quad)(\quad) = 4$$

Try to factor this one on your own. What factors make up x^2? Put your answers in the parentheses. What two factors will equal 4, when added and when multiplied? There is only one answer. Is this the answer you got?

$$(x + 2)(x + 2) = 4$$

Good, but we're not done. (x + 2) times (x + 2) is the same thing as $(x + 2)^2$. Let's rewrite the problem that way.

$$(x + 2)^2 = 4$$

Do you know what to do next? The number one rule in Algebra, when you are trying to solve for x, is to get x by itself. The next step, to get x by itself, is to get rid of the exponent. The way to get rid of a square, is to get the square root. And the number two rule is that whatever you do to one side, you must do to the other. So, let's get the square root of both sides of the equal sign.

$$\sqrt{(x+2)^2} = \sqrt{4}$$

The square root of $(x + 2)^2$ is $x + 2$. And the square root of 4 is plus or minus 2.

$$x + 2 = \pm 2$$

Now all we have to do is get rid of this 2. I will subtract 2 from both sides.

$$x = -2 \pm 2$$

Since we have a "plus or minus two," we are going to have two possible answers. First, we will solve for x, assuming the "plus or minus two" is positive two.

$$x = 0$$

Now we will solve for x assuming it is -2.

$$x = -4$$

We have just solved for x in the quadratic equation below. We ended up with 2 possible answers. Try replacing the x's with each possible answer. You will see that they are both right.

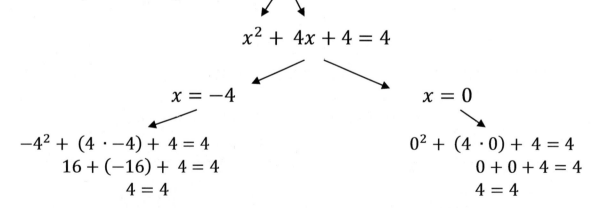

$$x^2 + 4x + 4 = 4$$

$x = -4$ 	$x = 0$

$-4^2 + (4 \cdot -4) + 4 = 4$	$0^2 + (4 \cdot 0) + 4 = 4$
$16 + (-16) + 4 = 4$	$0 + 0 + 4 = 4$
$4 = 4$	$4 = 4$

Yep, I can see that both answers are right because both sides of the equal sign are equal! Let's try another one together. Look at the next example.

$$x^2 + 12x + 20 = 0$$

The first step is to factor the quadratic. The first two terms are both x and the signs must be positive.

$$(x +)(x +) = 0$$

What two factors will equal 12 when added, and equal 20 when multiplied? I can only think of one possibility, 10 and 2.

$$(x + 10)(x + 2) = 0$$

Let's stop there for a moment and take a closer look at this equation. What is this problem saying? It is saying that (some number) times (some number) = zero. Mmm…in order for the answer to be zero, one of those numbers inside the parentheses must equal zero too, right? Because any number times zero equals zero.

So, if I want the answer to this problem to equal zero, all I have to do is make sure one of those binomials equals zero too. That way, the answer to this quadratic equation HAS to equal zero!

$$(x + 10)(x + 2) = 0$$

Remember, I want one of these two binomials to equal zero, so the equation is true. Look at the first binomial, (x + 10). If x = -10, then that binomial would equal 0, right? Look at the math to prove it. I say that x = -10.

$$(-10 + 10)(-10 + 2) = 0$$

$$0(-8) = 0$$

$$0 = 0$$

Once you prove that both sides of the equal sign are indeed equal, you know you have the right answer. However, look at the other binomial, (x + 2). I could make that binomial equal zero instead. If x = -2, the quadratic equation would still be true. Here is the math to prove it.

$$x = -2$$

$$(-2 + 10)(-2 + 2) = 0$$

$$(8)(0) = 0$$

That may have seemed complicated, but if you were to write it on paper, this is all you would have to write. Try to follow along and read this math, line by line.

$$x^2 + 12x + 20 = 0$$
$$(x + 10)(x + 2) = 0$$
$$x = -10 \quad or \quad x = -2$$

The most difficult part is factoring the quadratic, so if you can do that, this should be pretty easy for you.

The next worksheet will have several quadratic equations. Some of them might be in the wrong order. Rearrange them, so they fit into the standard form of a quadratic equation. Make sure you give both possible answers and don't forget to use the $\pm$ sign, when getting the answer to a square root problem.

If you have difficulties, read the last few lessons again. It will make more sense the second time around.

Name_____ Date _____

Worksheet 31

Solve for x in the following quadratic equations.

1. $x^2 + 10x + 16 = 0$

2. $x^2 + x - 12 = 0$

3. $x^2 + 12x + 36 = 64$

4. $a^2 + 15 = -8a$

5. $x^2 - 11x = -24$

6. $(x + 3)^2 = 49$

7. $y^2 = -3y - 2$

8. $25 = x^2 - 18x + 81$

LESSON 32: A PERFECT SQUARE

Do you remember when I told you that a binomial expression, times another binomial expression, will always equal a quadratic expression? That's because a binomial has 2 terms and 2 x 2 = 4 (quad). For example, look at the problem below.

$$(x - 4)(x + 2) = x^2 + 2x - 4x - 8$$

A binomial x A binomial = a Quadratic
(2 term expressions)

When a binomial is squared, as shown in the example below, the quadratic answer is called a *perfect square*. A perfect square is any number that has two identical factors. The two identical factors in the problem below are (x + 7) and (x + 7).

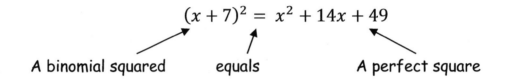

A binomial squared   equals   A perfect square

Here is an example of another perfect square that you might recognize.

$$49 = 7^2$$

The number 49 is a perfect square because it has two identical factors, 7 and 7. Here is another example.

$$x^2 = 25$$

You could do a little math and figure out that 25 is a perfect square too. It also has two identical factors, 5 and 5. But remember, in the equation above, there are actually TWO answers: "plus five" and "minus five." Since it is possible for x to be positive 5 or negative 5, we have to write the answer as ±5. That is read as plus or minus five. Here is another example. Can you solve for x?

$$x^2 = 49$$

In my head, I can figure out that the answer is x = 7, well...x = ±7. Let's look at the math.

$$x^2 = 49$$

Get the square root of each side.

$$\sqrt{x^2} = \sqrt{49}$$

The square root of any squared number is that number.

$$x = \pm 7$$

Can you solve for x in the equation below? I can't solve this one in my head, so let's write out the math, step by step.

$$(x-1)^2 = 16$$

To undo a square, get the square root of each side.

$$\sqrt{(x-1)^2} = \sqrt{16}$$

The square root of any number **squared** is that number.

$$x - 1 = \pm 4$$

Change the sign to + 1 and swing it to the other side.

$$x = +1 \pm 4$$

Do the math. There are two possible answers.

$$x = 5 \text{ or } x = -3$$

Do you understand why there are two possible answers? Since the square root of 16 can be plus OR minus 4, the answer has to consider both possibilities. If the square root of 16 is +4, then x = 5. If the square root of 16 is -4, then x = -3.

You can check to see if I got the right answers by putting my answers in place of the x in the original equation. I'll show you the math for both answers.

$$x = 5 \qquad\qquad x = -3$$
$$(x - 1)^2 = 16 \quad\text{original equation}\quad (x - 1)^2 = 16$$
$$(5 - 1)^2 = 16 \qquad\qquad (-3 - 1)^2 = 16$$
$$(4)^2 = 16 \qquad\qquad (-4)^2 = 16$$

Yep! Everything is equal, so those are the right answers.

When a quadratic equation is a perfect square, it is easy to solve for x. We will solve a few of them together. Here is an example of a perfect square. It might not look as obvious as the other examples, but trust me, it is a perfect square.

$$x^2 + 8x + 16 = 36$$

To solve for x in this quadratic equation, start by factoring the quadratic on the left side.

$$x^2 + 8x + 16 = 36$$
$$(x \quad)(x \quad) = 0$$

Can you guess which numbers to put inside the parentheses? They must equal 8 when added and 16 when they are multiplied. Mmm...this sounds like a job for 4. Wouldn't you agree?

$$x^2 + 8x + 16 = 36$$
$$(x \quad 4)(x \quad 4) = 36$$

Do you think we should put plus or minus signs inside the parentheses? Since the quadratic equation has only plus signs, it is safe to put a plus sign inside both sets of parentheses.

$$x^2 + 8x + 16 = 36$$
$$(x + 4)(x + 4) = 36$$

That looks like $(x + 4)^2$ to me, so let's rewrite it like that.

$$(x + 4)^2 = 36$$

To get rid of the square², find the square root of both sides.

$$\sqrt{(x + 4)^2} = \sqrt{36}$$

$$x + 4 = \pm 6$$

And finally, we get down to easy math. Swing the +4 to the other side and change the sign.

$$x = -4 \pm 6$$

Since it is possible for the square root of 36 to be either positive 6 or negative 6, we end up with two possible answers for x.

$$x = 2 \quad or \quad x = -10$$

OK, that was fun! Let's solve another perfectly square quadratic equation!

$$x^2 + 9 = 64 + 6x$$

Mmm...this one looks a little funny. Let's rearrange it, so it fits into our standard form of a quadratic equation. I will move the +6x to the other side, by changing the sign and putting it on the left side. I will put it in the middle, so it looks like the standard form: ax² + bx + c = 0.

$$x^2 - 6x + 9 = 64$$

Now we are ready to factor this side.

I've drawn an empty set of parentheses and filled in the first term of each with an x.

$$(x)(x) = 64$$

What two numbers will equal +9 when multiplied AND -6 when added? And of course, we are looking for a perfect square, so you know the two numbers will be the same. There is only one possible choice, -3. Fill in the parentheses with -3.

$$(x - 3)(x - 3) = 64$$

Rewrite that:
$$(x - 3)^2 = 64$$

Find the square root of each side:

$$\sqrt{(x-3)^2} = \sqrt{64}$$
$$x - 3 = \pm 8$$
$$x = 3 \pm 8$$
$$x = 11 \quad \text{or} \quad x = -5$$

See, it's not so bad. This next one will be even easier because some of the work has already been done.

$$(y + 4)^2 = 121$$

Do you know where to start? We are trying to solve for y, so start by getting the square root of both sides.

$$(y + 4)^2 = 121$$

$$\sqrt{(y+4)^2} = \sqrt{121}$$

$$y + 4 = \pm 11$$

$$y = -4 \pm 11$$

$$y = 7 \quad or \quad y = -15$$

See, when a quadratic equation is a perfect square, it is easy to solve for x. Many times, you will run into a problem like this next one.

$$x^2 + 6x + 9 = 30$$

Start by factoring the quadratic, as usual. This is an easy one, so I'll do it for you.

$$(x + 3)(x + 3) = 30$$

$$(x + 3)^2 = 30$$

$$\sqrt{(x + 3)^2} = \sqrt{30}$$

Let's stop here for a moment. I can find the square root of $(x + 3)^2$, that's easy, it's $(x + 3)$. But I'll need a calculator to find the square root of 30. Since the square root of 30 is 5.4772255575051661, you can just leave it at $\pm\sqrt{30}$. This is where we left off.

$$\sqrt{(x + 3)^2} = \sqrt{30}$$

$$x + 3 = \pm\sqrt{30}$$

$$x = -3 \pm \sqrt{30}$$

In this situation, you can just write the one answer. You don't have to say:

$$x = -3 + \sqrt{30} \quad or \quad x = -3 - \sqrt{30}$$

Try some yourself on the next worksheet.

Name_____ Date _____

Worksheet 32

Solve for x in the following quadratic equations.

1. $x^2 + 6x + 9 = 16$

2. $x^2 + 8x + 16 = 121$

3. $x^2 - 12x + 36 = 10$

4. $a^2 - 4a + 4 = 81$

5. $y^2 + 10y + 25 = 8$

6. $x^2 - 16x + 64 = 12$

7. $x^2 + 14x + 49 = 64$

8. $b^2 - 18b + 81 = 144$

LESSON 33: COMPLETING THE SQUARE

In that last chapter, you solved quadratic equations that were in the form of a perfect square. Unfortunately, life isn't always perfect. Often times, quadratic equations aren't in the form of a perfect square, which makes it a little more difficult to solve for x. But have no fear. Mathematicians have come up with a fancy little trick called *Completing the Square*, to help us.

First, I will show you the long way and then I will show you the short cut, by using the *Quadratic Formula*.

This math skill is called *Completing the Square*. It sounds kind of complicated, but once you get the hang of it, it's actually pretty easy.

Here is an example of a quadratic equation that is not a perfect square. Go ahead and try to factor this one. Draw some empty sets of parentheses and find the factors. You will soon see that they are NOT identical factors.

$$x^2 + 7x + 12 = 72$$

$$(x\ \ )(x\ \ ) = 72$$

Well let's see...I'm looking for two numbers that will equal 12 when multiplied and will equal 7 when added. I think 3 and 4 will work, let's try them out.

$$x^2 + 7x + 12 = 72$$

$$(x + 3)(x + 4) = 72$$

OK, we have successfully factored our quadratic expression, but we haven't solved for x yet. Looking at the two factors, (x + 3) and (x + 4), I can see that 72 is not a perfect square. We are going to have to use our "Completing the Square" trick, so we can force this equation to become a perfect square, making it easy to solve for x.

Let's start over from the beginning, before we factored our quadratic equation.

$$x^2 + 7x + 12 = 72$$

The very first thing you must do, when completing the square, is get rid of "c." That's the number that is throwing off our perfect square, so subtract it from both sides.

$$x^2 + 7x + 12 - 12 = 72 - 12$$

$$x^2 + 7x = 60$$

Now that we have gotten rid of "c," we can replace it with our own *special number*. The formula below will give us the special number we need to put in place of "c" to make it a perfect square.

$$\left(\frac{b}{2}\right)^2$$

This formula wants us to put the "b" from our equation in the place of this "b," do the math and then add that amount to each side of the equal sign.

Our equation: $x^2 + 7x + 12 = 72$
abc

The "b" in our equation is 7. Let's fill in the formula with the number 7 and then add that to each side.

$$\left(\frac{7}{2}\right)^2 \qquad b = 7$$

I could look at this as $(7 \div 2)^2$, but that would give us a decimal number. In algebra, we tend to shy away from decimal numbers and use fractions instead. So, I will look at this one as a fraction squared. The rule to squaring a fraction is to square both the numerator and the denominator separately.

$$\left(\frac{b}{2}\right)^2 = \frac{7^2}{2^2} = \frac{49}{4}$$

That answer is the special number that we will add to both sides of our equation. Don't worry about converting the improper fraction into a mixed number. In algebra, they leave fractions improper all the time.

You may be wondering, how we can just add a number to both sides of this equation. How is that possible? I can show you. Below is a scale with an equation on it. An equation is EQUAL on both sides of the equal sign – so the scale won't tip.

$$x^2 + 7x = 60$$

If I added 7xy to one side of the equal sign, the scale would tip because the two sides are no longer equal.

$$x^2 + 7x + 7xy = 60$$

But, if I add 7xy to BOTH sides, the equation is still equal!

$$x^2 + 7x + 7xy = 60 + 7xy$$

That's why it is OK to add a number to both sides. Back to our problem. This is how we left it. Let's add up the numbers on this side. Turn 60 into a fraction with a denominator of 4.

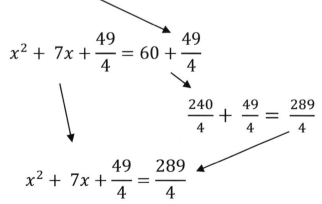

169

Now we are ready to factor our quadratic. Do you see how this equation is still in the standard form of a quadratic equation?

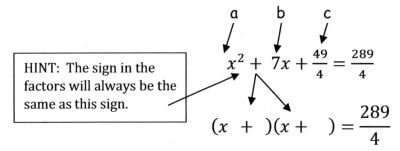

HINT: The sign in the factors will always be the same as this sign.

Next, we need two numbers that equal $\frac{49}{4}$ when multiplied and equal 7 when added. WHAT??? Are you kidding me? Don't worry; the answer is on the last page. We had to square $\frac{7}{2}$ to get $\frac{49}{4}$, so the answers are $\frac{7}{2}$ and $\frac{7}{2}$.

$$\left(x + \frac{7}{2}\right)\left(x + \frac{7}{2}\right) = \frac{289}{4}$$

Rewrite that like this; it's the same thing.

$$\left(x + \frac{7}{2}\right)^2 = \frac{289}{4}$$

Get the square root of both sides.

$$\sqrt{\left(x + \frac{7}{2}\right)^2} = \sqrt{\frac{289}{4}}$$

$$x + \frac{7}{2} = \pm \frac{17}{2}$$

$\frac{\sqrt{289}}{\sqrt{4}} = \frac{17}{2}$

$$x = \pm \frac{17}{2} - \frac{7}{2}$$

$$x = -\frac{24}{2} = -12 \quad or \quad x = \frac{10}{2} = 5$$

Let's review what we just did. We used $\left(\frac{b}{2}\right)^2$ to get a magical number to add to both sides of the equal sign. Doing this forced our quadratic equation into ending up with "something squared equals a number," making it easy for us to solve for x.

If you are a little confused, that's to be expected. This is pretty deep math. We'll go over this a couple more times. Here is our next example.

$$x^2 + 12x + 14 = 2$$

Look at this equation. If you try to factor this quadratic, you will quickly see that it is not a perfect square. No problem, we have a trick for that, it's the Completing the Square trick. I have matched up each letter from the standard form of a quadratic equation with each number from our quadratic equation.

$$ax^2 + bx + c = 0$$

$$1x^2 + 12x + 14 = 2$$

It's invisible

$$a = 1, \quad b = 12, \quad c = 14$$

Now that you know which numbers belong to a, b, and c; let's start the math. The first step is to get rid of "c." That is the number throwing off our equation. So, let's move the "c," or the "constant number," or "14," to the other side of the equal sign, by subtracting it from both sides. Here is the math.

$$x^2 + 12x + 14 = 2$$
$$x^2 + 12x = 2 - 14$$
$$x^2 + 12x = -12$$

Now it is time to complete the square. Let's fill in $\left(\frac{b}{2}\right)^2$ with the "b" from our problem.

$$b = 12$$

$\left(\frac{12}{2}\right)^2$ What is 12 ÷ 2?

$(6)^2$ What is 6^2?

The answer is 36. That is the magical number that we will add to both sides to force our quadratic equation into a perfect square. Let's give it a try.

We left off with:
$$x^2 + 12x = -12$$

Add 36 to both sides:
$$x^2 + 12x + 36 = -12 + 36$$

Do the math on the right side:
$$x^2 + 12x + 36 = 24$$

Factor the left side:
$$(x +)(x +) = 24$$

What two factors of 36 equal 12, when added together? I'll give you a hint: they will both be the same number. Of course, the number is 6.

Here is our problem now:
$$(x + 6)(x + 6) = 24$$

I'll rewrite that like this:
$$(x + 6)^2 = 24$$

Undo the square with a square root.
$$\sqrt{x + 6^2} = \sqrt{24}$$

Do the math.
$$x + 6 = \pm\sqrt{24}$$

Remember, the square root of 24 can be a negative number or a positive number; we don't know. Either way the answer will come out positive, so we HAVE TO say "plus or minus the square root of twenty-four" because it could be negative or positive.

Back to our problem…we left it at:
$$x + 6 = \pm\sqrt{24}$$

Subtract 6 from both sides:
$$x = -6 \pm \sqrt{24}$$

This problem is complete. We completed the square of a quadratic equation, making it possible to solve for x. Let's try another one.

$$x^2 - 2x = 10$$

This one looks a little different because there is no number for "c". No number is the same thing as 0, so if I put "+ 0" in place of the "c" in this equation, it would fit into our standard form of a quadratic equation.

$$x^2 - 2x + 0 = 10$$

But that is kind of silly because it means nothing, so let's take it back out. We are now back to our original problem.

$$x^2 - 2x = 10$$

We cannot factor the left side, so we are going to have to add a number to magically make it a perfect square. Do you remember the formula we used for our b term? Do you know which term is our b term in this problem? If the zero I added was "c," then "b" must be -2 (don't forget the negative sign).

$$\left(\frac{b}{2}\right)^2 = \left(\frac{-2}{2}\right)^2$$

Use your negative and positive skills to solve this one. What is -2 ÷ 2? I come up with -1. What is (-1)²? Let's see, -1 x -1 = 1, so that is our magical number for this problem. I will add 1 to both sides of the equal sign.

$$x^2 - 2x + 1 = 10 + 1$$

$$x^2 - 2x + 1 = 10 + 1$$

Factor the left side of the equal sign and do the math on the right.

$$(x\ \ )(x\ \ ) = 11$$

What goes inside the parentheses? We need two numbers that will equal 1 when multiplied, and will equal -2, when added. (Another clue is that they will be the same number). I think -1 will work, let's fill in the parentheses. Besides, -1 is the number we squared, so I know that's the number that goes inside the parentheses.

Here is our problem so far:
$$(x - 1)(x - 1) = 11$$

Rewrite that same problem:
$$(x - 1)^2 = 11$$

Get the square root of both sides:
$$\sqrt{(x - 1)^2} = \sqrt{11}$$

What is the square root of 11? It's 3.3166247903554. Sheesh! Let's just put down the calculator and write plus or minus the square root of eleven, instead.

$$x - 1 = \pm\sqrt{11}$$

Move the – 1 over and change the sign:

$$x = 1 \pm \sqrt{11}$$

We have just solved for x in a quadratic equation, by completing the square. Are you getting the hang of this yet? If not, read those last examples again. If you are catching on…good! Let's try a harder one.

$$x^2 - 4x - 3 = \frac{21}{25}$$

Yikes! This one looks rough. But don't worry you have all the math skills you need to solve for x in this quadratic equation. We will start by moving the -3 (c) over to the other side.

$$x^2 - 4x = \frac{21}{25} + 3$$

We need a common denominator to add fractions. I will use 25 for a common denominator. Here is the math.

$$\frac{21}{25} + \frac{75}{25} = \frac{96}{25}$$

$$x^2 - 4x = \frac{96}{25}$$

It is time to use our completing the square trick. Plug the "b" term into the formula. Remember, it is a negative 4!

$$\left(\frac{b}{2}\right)^2 \longrightarrow \left(\frac{-4}{2}\right)^2 = (-2)^2 = 4$$

What is your answer? I get 4. Do you know what to do with this 4? Add it to both sides. This is how it should look now.

$$x^2 - 4x + 4 = \frac{96}{25} + 4$$

Now let's factor the left side and add the right side.

$$x^2 - 4x + 4 = \frac{96}{25} + 4$$

$$(x \quad)(x \quad) = \frac{96}{25} + \frac{100}{25}$$

What two factors go in the parentheses? Here are your clues:
- They will both be the same number
- They will equal -4, when added together
- They will equal 4, when multiplied together

I can only think of one possible answer, -2.

$$(x-2)(x-2) = \frac{196}{25}$$

Rewrite it:
$$(x-2)^2 = \frac{196}{25}$$

Get the square root of each side.

$$\sqrt{(x-2)^2} = \sqrt{\frac{196}{25}}$$

$$\frac{\sqrt{196}}{\sqrt{25}} = \pm\frac{14}{5}$$

$$x - 2 = \pm\frac{14}{5}$$

Get rid of that -2, by adding 2 to both sides. We have now solved for x, however, there is still some math that can be done. I will solve to get the two possible answers for x.

$$x = 2 \pm \frac{14}{5}$$

$$x = \frac{24}{5} \quad or \quad x = -\frac{4}{5}$$

Let's go over one more example together. This time, our "a" won't be invisible. When there is a coefficient, besides one, in the first term, you have to divide it out. You can try to solve this problem by completing the square, but you will soon get stuck when you have to find two identical factors of 2x.

$$2x^2 + 6x - 12 = 2$$

To get rid of this "2," you have to divide the whole equation by 2. The trick here is to divide EACH term by 2.

$$\frac{\overset{1}{\cancel{2}}x^2 + \overset{3}{\cancel{6}}x - \overset{6}{\cancel{12}} = \overset{1}{\cancel{2}}}{\underset{1}{\cancel{2}}}$$

Now our quadratic equation looks like this:

$$x^2 + 3x - 6 = 1$$

Let's try to solve for x by completing the square. The first step is to get rid of "c." Add 6 to both sides.

$$x^2 + 3x - 6 + 6 = 1 + 6$$

$$x^2 + 3x = 7$$

Next, find the magical number.

$$\left(\frac{3}{2}\right)^2 = \left(\frac{9}{4}\right)$$

Add it to both sides.

$$x^2 + 3x + \frac{9}{4} = 7 + \frac{9}{4}$$

Remember, this number can be found in our previous work.

Factor the left side and add the right side.

$$\left(x + \frac{3}{2}\right)\left(x + \frac{3}{2}\right) = \frac{37}{4}$$

Rewrite it:
$$\left(x+\frac{3}{2}\right)^2 = \frac{37}{4}$$

Get the square root of each side.

$$\sqrt{\left(x+\frac{3}{2}\right)^2} = \sqrt{\frac{37}{4}}$$

$$x+\frac{3}{2} = \frac{\sqrt{37}}{2}$$

Get x by itself.

$$x = -\frac{3}{2} \pm \frac{\sqrt{37}}{2}$$

Let's try another one together. You can try to factor this one if you want, but I don't think you'll get very far. Go ahead and try. I'll wait…are you ready to try to complete the square now?

I will start by circling the "b" term and the "c" term because those are the two terms we need to focus on.

$$x^2 \,\textcircled{+ 6x} \,\textcircled{-\frac{4}{3}} = 15$$

Since this quadratic is a real brain teaser to factor, we are going to complete the square instead.

The first step is to move "c" over to the other side.

This is done by adding it to both sides.

$$x^2 + 6x = 15 + \frac{4}{3}$$

$$\frac{45}{3} + \frac{4}{3} = \frac{49}{3}$$

$$x^2 + 6x = \frac{49}{3}$$

Next, we need to use the formula that will give us a perfect little number to make our equation a perfect square.

$$\left(\frac{b}{2}\right)^2 \longrightarrow \left(\frac{6}{2}\right)^2 = (3)^2 = 9$$

The formula has given us the number 9 to add to both sides of our equation.

$$x^2 + 6x + 9 = \frac{49}{3} + 9$$

$$\frac{49}{3} + \frac{27}{3} = \frac{76}{3}$$

$$x^2 + 6x + 9 = \frac{76}{3}$$

$$(x +)(x +) = \frac{76}{3}$$

Next, we need to find two identical numbers that will equal 9 when multiplied, and will equal 6 when added. That's easy, 3 and 3. Try to solve the rest of the problem on your own and then look at my answer.

$$(x + 3)(x + 3) = \frac{76}{3}$$

$$(x + 3)^2 = \frac{76}{3}$$

$$\sqrt{(x+3)^2} = \sqrt{\frac{76}{3}}$$

$$x + 3 = \pm \frac{\sqrt{76}}{\sqrt{3}}$$

$$x = -3 \pm \frac{\sqrt{76}}{\sqrt{3}}$$

This problem is solved. Try to solve a few on your own on the next worksheet. Remember the steps:
- Get rid of "c".
- Use $\left(\frac{b}{2}\right)^2$ to get a special number.
- Put that special number on both sides of the equal sign.
- Solve the math.

Name_____ Date _____

Worksheet 33

Solve the following quadratic equations, by completing the square with $\left(\frac{b}{2}\right)^2$.

1. $x^2 - 2x - 2 = 0$

2. $x^2 - 2x - 10 = 0$

3. $x^2 + 4x - 15 = 0$

4. $4x^2 - 8x - 24 = 4$

HINT: If there is a coefficient in the "a" position, you must divide it out before trying to complete the square.

LESSON 34: THE QUADRATIC FORMULA

That last lesson was the long way of solving quadratic equations. Luckily, a mathematician came up with a formula for us to use instead. This is the third way to solve a quadratic equation. It is called the Quadratic Formula, and this is what it looks like.

$$x = \frac{-b \pm \sqrt{b^2 - 4ac}}{2a}$$

I know it looks a little scary, but it's a lot easier than completing the square. Just fill in a, b, and c with the numbers from any quadratic equation and you can solve for x without using the Completing the Square trick. Let's give one a try.

Below is a quadratic equation. We are going to solve for x by using the Quadratic Formula, written above. First, I will identify each term as either a, b, or c.

$$x^2 + 4x - 8 = 4$$

$$\uparrow \quad \uparrow \quad \uparrow \qquad ?$$
$$a \quad \; b \quad \; c$$

What do we do with the 4 at the end? We need to get rid of it. Our equation must look just like the standard form of a quadratic equation. That means our equation must "equal zero," so minus 4 from both sides.

$$x^2 + 4x - 8 - 4 = 4 - 4$$
$$x^2 + 4x - 12 = 0$$

Notice how our "c" is now -12. I will list the value of each variable below. Now we are ready to put our numbers into the quadratic formula.

a = 1 b = 4 c = -12

$$x = \frac{-4 \pm \sqrt{4^2 - 4 \cdot 1 \cdot -12}}{2 \cdot 1}$$

Next, we have some math to solve. Let's start with the stuff under the square root sign.

$$x = \frac{-4 \pm \sqrt{4^2 - (4 \cdot 1 \cdot -12)}}{2 \cdot 1}$$

$$\sqrt{4^2 - (4 \cdot 1 \cdot -12)} =$$

$$\sqrt{16 - (-48)} =$$

$$x = \frac{-4 \pm \sqrt{64}}{2 \cdot 1}$$

Next, solve the math in the denominator and the square root.

$$x = \frac{-4 \pm 8}{2}$$

What are the two possible answers?

$$x = \frac{4}{2} = 2 \quad or \quad x = \frac{-12}{2} = -6$$

And that's all there is to it! You can check my work by putting my answers in place of "x" in the original equation. See, it's not so bad; it just looks complicated. Let's try another one together.

Below, is a quadratic equation. We will use the Quadratic Formula to solve for x. You can see "a, b, and c" below, but there is one more number that doesn't have a letter, $\frac{21}{25}$.

$$x^2 - 4x - 3 = \frac{21}{25}$$

a = 1   b = -4   c = -3

In order to use the Quadratic Formula, we must get our equation into the standard form of a Quadratic Equation. That means we need our equation to equal zero, not

$\frac{21}{25}$. No problem, let's just subtract it from both sides; that will keep our scale from tipping.

$$x^2 - 4x - 3 - \frac{21}{25} = \frac{21}{25} - \frac{21}{25}$$

$$x^2 - 4x - \frac{75}{25} - \frac{21}{25} = 0$$

$$x^2 - 4x - \frac{96}{25} = 0$$

Now it looks like the standard form of a quadratic equation, so we are ready to fill in the quadratic formula with our numbers:

$$x^2 - 4x - \frac{96}{25} = 0$$

$$a = 1 \quad b = -4 \quad c = -\frac{96}{25}$$

Quadratic Formula

$$x = \frac{-b \pm \sqrt{b^2 - 4ac}}{2a}$$

$$x = \frac{-(-4) \pm \sqrt{(-4)^2 - 4 \cdot 1 \cdot \left(-\frac{96}{25}\right)}}{2(1)}$$

Let's simplify this problem, by solving the multiplication.

$$-\frac{4}{1} \cdot -\frac{96}{25} = +\frac{384}{25}$$

$$x = \frac{+4 \pm \sqrt{16 + \frac{384}{25}}}{2}$$

Next, solve the math under the square root sign.

$$\frac{400}{25} + \frac{384}{25} = \frac{784}{25}$$

$$x = \frac{4 \pm \sqrt{\frac{784}{25}}}{2}$$

$$x = \frac{4 \pm \sqrt{\frac{784}{25}}}{2}$$

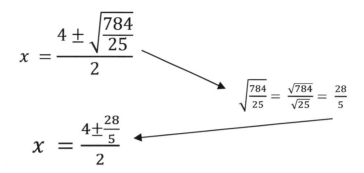

$$\sqrt{\frac{784}{25}} = \frac{\sqrt{784}}{\sqrt{25}} = \frac{28}{5}$$

$$x = \frac{4 \pm \frac{28}{5}}{2}$$

This means the answer to x is either:

$$x = \frac{\frac{20}{5} + \frac{28}{5}}{2} \quad or \quad x = \frac{\frac{20}{5} - \frac{28}{5}}{2}$$

I'll solve the math.

$$\frac{48}{5} \div 2 = \quad \frac{48}{5} \cdot \frac{1}{2} = \frac{48}{10} = \frac{24}{5} \qquad -\frac{8}{5} \div \frac{2}{1} = \quad -\frac{8}{5} \cdot \frac{1}{2} = -\frac{8}{10} = -\frac{4}{5}$$

In the end, we get the same answer that we did when we used the Completing the Square trick.

$$x = \frac{24}{5} \quad or \quad x = -\frac{4}{5}$$

Try solving some quadratic equations using the Quadratic Formula on the next worksheet.

Name_____ Date _____

Worksheet 34

Use the Quadratic Formula to solve the following quadratic equations.

$$x = \frac{-b \pm \sqrt{b^2 - 4ac}}{2a}$$

1. $3x^2 - 2x - 5 = 0$

2. $2x^2 - 5x + 3 = 0$

3. $y^2 + 2y - 8 = 0$

LESSON 35: INTRO TO FUNCTIONS

Do you remember back in pre-algebra, when you learned slopes? If you don't understand slopes, read Volume III of the *Learn Math Fast System*. During the lesson on slopes, you were given a linear equation and then you were told to solve for x in terms of y. When you did this, you ended up with some coordinates that you then plotted on a graph. The linear equation ended up creating a straight line on the graph.

Now we are working with a *quadratic equation*, instead of a linear equation. A linear equation draws a line on a graph, but a quadratic equation draws a parabola (pa-rab'-o-la). A parabola looks like the u-shape drawn on this graph.

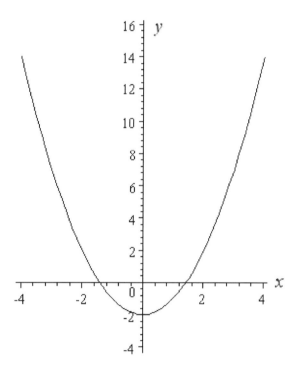

When we are given a quadratic equation, like the one below, we solve for x and end up with two answers.

$$x^2 - 2x - 3 = 0$$

I will solve for x and get the two possible answers.

$$x^2 - 2x - 3 = 0$$

$$(x - 3)(x + 1) = 0$$

$$x = 3 \quad or \quad x = -1$$

These two answers are the coordinate points on the x axis. Having our equation equal to zero, means the y coordinates are 0. If I replace the zero with the letter y, I create the *standard form of a quadratic function*:

$$x^2 - 2x - 3 = y$$

Now I can assign any number to x and I will get some coordinates for y. The standard form of a quadratic function is written like this:

$$ax^2 + bx + c = f(x)$$

At the top of this page, I solved a quadratic equation and I got two answers for x. Now I will use a *quadratic function* to find more coordinates, so we can graph a parabola. This is called graphing a function. Here are the answers I have from the equation at the top of the page.

x	f(x)
-1	0
3	0

This time I will use a quadratic function. The answer to this equation will give us a "y" coordinate. When you use a function, the variable x is assigned a number. You are to replace each x with that number. Below is a quadratic function.

$$f(x) = x^2 - 2x - 3$$

I will assign x to be -2. Now the function looks like the one below. Notice how every x has been turned into a -2.

$$f(-2) = (-2)^2 - 2(-2) - 3$$

This equation is read as "The f of $-2 = (-2)^2 - 2(-2) - 3$." When a function says the $f(-2)$, it means replace all the x's with a -2. That's what I've done here:

$$f(-2) = (-2)^2 - 2(-2) - 3$$

$$f(-2) = 4 + 4 - 3$$

$$f(-2) = 5$$

This means that if our x coordinate equals -2, then our y coordinate will equal 5. I will add that answer to our table.

x	f(x)
-1	0
3	0
-2	5

Let's get a few more points. Then we will graph the points and draw our parabola. This next equation will be $f(4)$. That means you should replace each x with the number 4.

$$f(x) = x^2 - 2x - 3$$

$$f(4) = 4^2 - 2(4) - 3$$

$$f(4) = 16 - 8 - 3$$

$$f(4) = 5$$

Add that to our table:

x	f(x)
-1	0
3	0
-2	5
4	5

One more time, graph the function. This time we will have the $f(1)$.

$$x^2 - 2x - 3 = f(1)$$

$$(1)^2 - 2(1) - 3 = f(1)$$

$$1 - 2 - 3 = f(1)$$

$$-4 = f(1)$$

Add that to our table.

x	f(x)
-1	0
3	0
-2	5
4	5
1	-4

Now we will plot the points and draw our parabola.

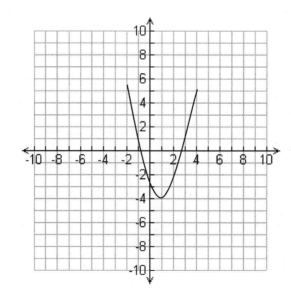

Mathematicians use quadratic equations to draw parabolas on graphs. They can derive important information from these parabolas. When you are given a quadratic equation and then you find the coordinates to draw a parabola, you are *graphing the function.*

On the next worksheet, you will be given a quadratic function. Solve each equation by replacing the x's.

Name_____ Date _____

Worksheet 35

1. Find $f(5)$ in the following quadratic function.
   $f(x) = x^2 + 2x - 35$

2. Find $f(-2)$ in the following quadratic function.
   $f(x) = x^2 - 8x - 20$

3. Find $f(8)$ in the following quadratic function.
   $f(x) = x^2 - 3x - 30$

4. Find $f(3)$ in the following quadratic function.
   $x^2 + 4x - 10 = f(x)$

5. Find $f(4)$ in the following quadratic function.
   $f(x) = x^2 + 4x - 22$

6. Find $f(6)$ in the following quadratic function.
   $f(x) = x^2 + \frac{1}{2}x - \sqrt{64}$

7. Find $f(7)$ in the following quadratic function.
   $f(x) = x^2 + 9x - 84$

LESSON 36: FINAL REVIEW

OK, so the beginning of the book was nice and easy, but once we got to the polynomials and quadratics, it got a little more confusing. Leaving the math behind, let me explain quadratics logically.

First, you were taught about binomials. Then you were taught how to multiply two binomials together, using the FOIL method. And what is the product of two binomials called? That's right - a quadratic; just like the one written below.

$$x^2 + 3x + 6$$

Above is a quadratic, but once I add "equals something" to it, it becomes a *quadratic equation*. Below is an example of a quadratic equation.

$$x^2 + 3x + 6 = 0$$

This equation is equal to zero. When a quadratic equation is set to equal zero, solving for x will give you two x coordinates. The two y coordinates are 0. These two sets of coordinates show you where to draw two points on a graph. This is very similar to slopes, which you should have learned about in pre-algebra. If you don't understand slopes, read Volume III of the *Learn Math Fast System*. You can learn about them in just a day or two.

Sometimes, it is difficult to solve for x in a quadratic equation, so we have a couple tricks we can use. The first trick is "completing the square." We use the formula $\left(\frac{b}{2}\right)^2$ to find a special number to add to both sides. BUT, before you can do that, you must move your "c" term over to the other side of the equal sign. We do this because the "c" number is the reason it is impossible to factor.

If that trick doesn't work for you, there is a third way. You can use the quadratic formula; it is written below.

$$x = \frac{-b \pm \sqrt{b^2 - 4ac}}{2a}$$

Before you can use the quadratic formula, you must make sure your equation is equal to zero. We do this because the quadratic formula only has reserved seating for "a, b, and c." We can't leave a number behind, so you must include it with "c."

Those are the three ways to solve a quadratic equation. No matter which way you choose, you will always end up with two possible answers for x. Why are there always two answers, you ask? Well...picture a parabola, the left and right sides are a mirror image of each other. So, each quadratic equation gives a place for x on the left, and a place for x on the right. Both will have the same y coordinate.

How do I know the "y" coordinate, you ask? Well...if the equation is equal to zero, then that is your y coordinate. BUT, if you have a quadratic function, like the one below, your "y" coordinate is the answer to $f(x)$. And your x coordinate is the one in the parentheses, for example, $f(3)$.

$$x^2 + 4x + 4 = f(x)$$

Maybe it would make more sense to you, if I wrote the same quadratic function like this:

$$x^2 + 4x + 4 = y \qquad x = 3$$

That means the same thing as the quadratic function above. It means: If x = 3, what will the y coordinate be, according to this equation? And remember, a quadratic equation is used to draw a parabola on a graph. You will learn how to use the information you get from a parabola, in Algebra II.

Ultimately, we are solving these equations and functions to get a bunch of coordinates to plot on a graph. Once you have around eight points drawn on your graph, connect the dots, and you should have drawn a parabola. If it doesn't, then you have a wrong answer somewhere.

On the final test, you will solve quadratic equations, using all three methods. Then you will solve quadratic functions to gather a handful of coordinates that you will plot on a graph. Good Luck.

Name_____ Date _____

FINAL TEST

Solve the following quadratics by factoring. Check your answers, by replacing x with your answer.

1. $x^2 - 2x - 24 = 0$

2. $x^2 - 3x - 28 = 0$

3. $x^2 + 22x + 121 = 0$

4. $x^2 - x - 6 = 0$

Continued on next page...

Final Test page 2

Solve the following quadratics by completing the square. $\left(\frac{b}{2}\right)^2$

5. $x^2 - \frac{1}{2}x - 3 = 0$

6. $x^2 + 10x - 24 = 0$

Solve the following quadratics by using the Quadratic Formula. $x = \frac{-b \pm \sqrt{b^2 - 4ac}}{2a}$

7. $2x^2 + 5x = 12$

8. $x^2 + 9x + 4 = 4$

Continued on next page...

Final Test page 3

Below is a quadratic function. Solve for each $f(x)$. Put your answers on the table. Graph all the coordinate points you create, on the next page of this test. Then draw the parabola you created with this quadratic function.

$$x^2 - 4x - 12 = 0$$

9. First solve for x in the equation above. Put your two answers on the table next to the 0's.

10. $f(1)$

11. $f(2)$

12. $f(-1)$

13. $f(4)$

14. $f(5)$

15. $f(3)$

x	f(x)
	0
	0

Continued on next page…

Final Test page 4

Plot the points from the table on the last page.

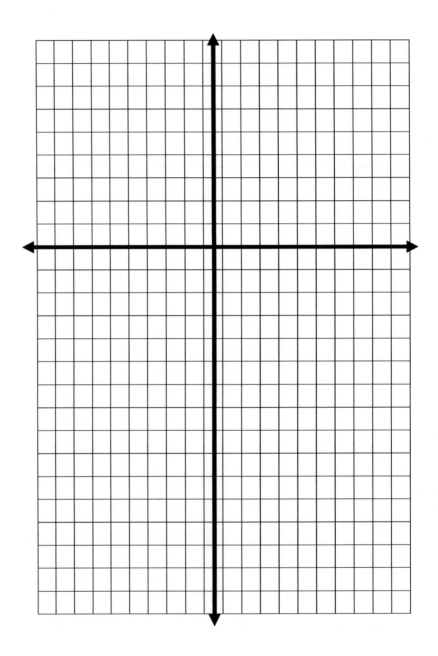

ANSWERS

The following pages contain the answers to every worksheet and test. Compare your answers to these answers to make sure you are getting all the problems correct. If you get stuck on a problem, use these answers to help you learn.

ANSWERS: Pre-Algebra Review Test

1. $-7 + -7 = -14$

2. $5 + -8 = -3$

3. $(7) - (-3) = 10$

4. $6 + -10 = -4$

5. $-\frac{3}{8} + \frac{7}{24} = -\frac{2}{24}$ or $-\frac{1}{12}$

6. $\frac{2}{7} - -\frac{5}{21} = \frac{11}{21}$

7. $-10 - 6 = -16$

8. $7 \times -6 = -42$

9. $-5 \times -5 = 25$

10. $15 \div -3 = -5$

11. $-24 \div -8 = 3$

12. $-48 \div 6 = -8$

13. $-\frac{1}{2} \times 2 = -1$

14. $-.07 \times -.7 = .049$

15. $-9.3 \cdot 4.1 = -38.13$

16. $\frac{3}{4}\left(-\frac{5}{8}\right) = -\frac{15}{32}$

17. $\frac{1}{3} \div \frac{1}{2} = \frac{1}{3} \times \frac{2}{1} = \frac{2}{3}$

18. $-\frac{4}{8} \div \frac{3}{4} = -\frac{16}{24}$ or $-\frac{2}{3}$

19. $\begin{aligned} 10 + x &= 13 \\ -10 & -10 \\ \hline x &= 3 \end{aligned}$

20. $\begin{aligned} x - 10 &= 90 \\ +10 & +10 \\ \hline x &= 100 \end{aligned}$

21. $\begin{aligned} 5x &= 30 \\ 5x \div 5 &= 30 \div 5 \\ x &= 6 \end{aligned}$

22. $\begin{aligned} \frac{55}{x} &= 5 \\ x\left(\frac{55}{x}\right) &= 5x \\ 55 &= 5x \\ 11 &= x \end{aligned}$

23. $\begin{aligned} \sqrt{64} &= x \\ 8 &= x \end{aligned}$

24. $\begin{aligned} 3^2 \div 3 + 6 &= x \\ (9 \div 3) + 6 &= x \\ 3 + 6 &= x \\ 9 &= x \end{aligned}$

25. $\begin{aligned} x - 1 &= y \\ +1 & +1 \\ \hline x &= y + 1 \end{aligned}$

26. $\begin{aligned} x + 2 &= y \\ -2 & -2 \\ \hline x &= y - 2 \end{aligned}$

27. $\begin{aligned} 5 + x &= y \\ -5 & -5 \\ \hline x &= y - 5 \end{aligned}$

28. Use the slope formula to find the slope of a line with coordinates (2, 3) and (3, 5).

$$m = \frac{5 - 3}{3 - 2} = \frac{2}{1} = 2$$

ANSWERS: Worksheet 1

1. 5, 7, 18, 29
   All Natural Numbers
   (and integers)

2. -2, 1, 3, 9
   All Integers

3. 0, 1, 2, 3
   All Integers

4. .5, 1, 1.5, 2, 2.5
   Neither

5. -20, -21, -22
   All Integers

6. 1015, 1025, 1046
   All Natural Numbers
   (and integers)

7. 1, ½, 3, 5
   Neither

8. -1, 0, 1
   All Integers

9. 253.5, 600.5, 3
   Neither

10. 7, 12, -3, 0, 5
    All Integers

11. 14, 12, 110, 6
    All Natural Numbers

12. ½, .335, $\sqrt{37}$, 0
    Neither

13. In t eg ers
    Includes the negative numbers and zero

14. Natural Numbers

15. Real Numbers

16. $-3 \times -9 = 27$

17. $\frac{1}{4} \div \frac{3}{5} =$
    $\frac{1}{4} \times \frac{5}{3} = \frac{5}{12}$

18. $10 + x = -19$
    $\underline{-10 \quad\quad -10}$
    $x = -29$

19. $\frac{49}{7} = 7$

20. $3a = 99$
    $3a \div 3 = 99 \div 3$
    $a = 33$

21. $\frac{1}{2} \cdot \frac{1}{2} = \frac{1}{4}$

22. $\sqrt{121} = 11$

23. $\frac{1}{2}x = 32$
    $\frac{1}{2}x \div \frac{1}{2} = 32 \div \frac{1}{2}$
    $x = 64$

24. $12^2 = 144$

25. $3\frac{4}{7} \times 4\frac{3}{5} = \frac{25}{7} \times \frac{23}{5} = \frac{575}{35} = 16\frac{15}{35} = 16\frac{3}{7}$

26. $4^2 + (3 - 1) + 5(8 + 2) =$
    $4^2 + (2) + 5(10) =$
    $16 + 2 + 50 = 68$

ANSWERS: Worksheet 2

*   17   A rational, natural number and an integer.
1.   -17   A rational integer.
2.   .34   A rational number that is neither an integer nor a natural number.
3.   .33   A rational number that is neither an integer nor a natural number.
4. $\sqrt{4}$ A rational, natural number and an integer.
5. $\sqrt{2}$ An irrational number that is neither an integer nor a natural number.
6.   -7/8   A rational number that is neither an integer nor a natural number.
7.   0   A rational integer.
8. π An irrational real number.
9.   100   A rational natural number and an integer.
10.   1/3   A rational number that is neither an integer nor a natural number.

11. $3a = 27$
    $3a \div 3 = 27 \div 3$
    $a = 9$

12. $\frac{1}{2}a = 5$
    $\frac{1}{2}a \div \frac{1}{2} = 5 \div \frac{1}{2}$
    $a = 10$

13. $-12\ x - 12 = 144$

14. $-8^2 = 64$

15. $\sqrt{36} = 6$

16. $x - 5 = -10$
    $ + 5 = +5$
    $x = -5$

17. $8(6 + 3) =$
    $8 \times 9 = 72$

18. $(3 + 3) - 2^2 + (4 \times 6)(\sqrt{4}) =$
    $(6) - 2^2 + (24)(2) =$
    $(6\ \ 4)\ |\ 48 = 50$

19. Parentheses, Exponents, Multiplication, Division, Addition, Subtraction

20. $-\frac{5}{9} \div \frac{5}{6} =$
    $-\frac{5}{9} \times \frac{6}{5} = -\frac{30}{45} = -\frac{2}{3}$

21. $\frac{7}{12} \times -\frac{8}{11} = -\frac{56}{132} = -\frac{14}{33}$

22. $2\frac{7}{9} + 4\frac{4}{27} = 6\frac{25}{27}$

ANSWERS: Worksheet 3

Name the variable in each problem.

1. $5a - 7 =$ a
2. $a + 2 =$ a
3. $5b =$ b
4. $10a + 4a =$ a
5. c   c

Name the coefficient of y in each problem.

6. $3y + 7 =$ 3
7. $4a - 2y =$ -2
8. $\frac{1}{2}y =$ $\frac{1}{2}$
9. $y + 1 =$ 1
10. $abc + 6y =$ 6

Look at this problem then answer the following questions about it.

$$-8b + 3 =$$

11. Is the coefficient of b an integer?   Yes
12. Is the coefficient of b a natural number?   No, it is negative.
13. Is the coefficient of b an irrational number?   No
14. Is the number without a variable a rational number? Yes
15. What is the variable?   b

16. $-\frac{4}{7} \times \frac{4}{7} = -\frac{16}{49}$

17. $-\frac{4}{7} \div \frac{4}{7} =$
    $-\frac{4}{7} \times \frac{7}{4} = -\frac{28}{28} = -1$

18. $-\frac{4}{7} - \frac{4}{7} = -\frac{8}{7}$
    $or -1\frac{1}{7}$

ANSWERS: Worksheet 4

Name each of the following as either an expression or a term.

1. 4y — **Term**
2. abc — **Term**
3. $12ab^2$ — **Term**
4. 3a + 4b — **Expression**
5. 3^2 - (5a + 2c) — **Expression**
6. $6y^2 + 8$ — **Expression**
7. 1/2ab(3a) — **Term**
8. 4 + 3xyz — **Expression**
9. 9a — **Term**
10. dπ — **Term**

Fill in the blanks with a word that matches the description.

11. **Integers** — A number set that includes all negative and positive whole numbers including zero.
12. **Natural Numbers** — The set of numbers that we use to count.
13. **Real Numbers** — The set of numbers that contain every number.
14. **Rational** — The type of numbers that can be written as a fraction.
15. **Irrational** — The type of numbers that cannot be written as a fraction.
16. **Expression** — Any number of terms hooked together with a plus or minus sign.
17. Are there any like terms in the problem below? If so, what are they?
    $7xy + abc + 4xy + 5abc + 7xy^2$ = **Yes**

          7xy and 4xy   and   **abc and 5abc.**

ANSWERS: Worksheet 5

	Term	Factors					
1.	$7y^3$	7	y	y	y		
2.	$3xy^2$	3	x	y	y		
3.	$5x$	5	x				
4.	a^4	a	a	a	a		
5.	$11xy^2z$	11	x	y	y	z	
6.	3^3	3	3	3			
7.	7 x 2	7	2				
8.	8 + 5	no factors					
9.	3a	3	a				
10.	$21a^3b^2$	7	3	a	a	a	b	b
11.	(3 + 2)(4 + b)	(3+2)	(4 + b)				
12.	10^2	10	10				
13.	3(a + b)	3	(a + b)				
14.	a(2 + a)	a	(2 + a)				
15.	ab	a	b				

Solve the following problems.

16. $|8| + |-9| = 17$

17. $|-3| + |5| = 8$

18. $|7| - |-3| = 4$

19. $|-12| - |-5| = 7$

20. $|-25| + |5| = 30$

21. $|-9| + |3| + |-6| = 18$

22. $|-5| \times |9| = 45$

23. $|7| \times |-8| = 56$

24. $|-4| \times |4| = 16$

ANSWERS: Chapter 1 Review Test

Name each number as either a natural number, an integer, both, or neither.

1. 5	2. -5	3. $\frac{1}{5}$	4. $-\frac{1}{5}$	5. 0
Both	**Integer**	**Neither**	**Neither**	**Integer**

Name each number as rational or irrational.

6. 3/8	7. $\sqrt{49}$	8. $\sqrt{2}$	9. π	10. 2
Rational	**Rational**	**Irrational**	**Irrational**	**Rational**

Name the variable and the coefficient of each term.

11. $2a$	12. b	13. $10x^2$	14. $\frac{5}{8}m$	15. πr^2
Var.= a	Var.= b	Var.= x	Var.= m	Var.= r
Coeff= 2	Coeff= 1	Coeff= 10	Coeff= 5/8	Coeff=3.14

Name each of the following as either a term or an expression.

16. $3xy^2$	17. $8a(3y)$	18. $\sqrt{9} + \sqrt{7}$	19. $6a + 12y$	20. a
Term	**Term**	**Expression**	**Expression**	**Term**

Look at the expression below. Circle all like terms.

21. $\underbrace{3a^2b^2}$ + 7ab + $\underbrace{a^2b^2}$ + $4ab^2$ – $7a$ – $\underbrace{6a^2b^2}$ + $5a^2b$ + $9a^2b^2c$ =

List all the factors of the following terms.

22. 5ab	23. $7y^2$	24. 11abc	25. $(5a + 3)(2a + 4)$	26. $(a + b)^2$
5, a, b	**7, y, y**	**11, a, b, c**	**(5a + 3), (2a + 4)**	**(a + b), (a + b)**

Solve the following.

27. $|-9|$ = **9**   28. $|8| - |-3|$ = **5**   29. $|27| * |-2|$ = **54**

ANSWERS: Worksheet 6

Add or subtract like terms.

1. $m^2 + m^2 = 2m^2$
2. $2a^3 + 3a^3 = 5a^3$
3. $4a^2b + 5a^2b = 9a^2b$
4. $6y^4 + 3y = 6y^4 + 3y$
5. $a^3 + a^3 + a^3 = 3a^3$
6. $3a^5 + 2a^5 + a^5 = 6a^5$
7. $7in^2 + 3in^2 = 10in^2$
8. $9ab^2 + 7a^2 = 9ab^2 + 7a^2$
9. $a^4 + 3a^4 = 4a^4$
10. $mn + 3mn^2 + 2mn^2 = mn + 5mn^2$

11. $5ab^3 - ab^3 = 4ab^3$
12. $9x^3 - 3x^3 = 6x^3$
13. $4y^2 - 2y^3 = 4y^2 - 2y^3$
14. $15xy^4 - 12xy^4 = 3xy^4$
15. $11a^5 - 9a^6 = 11a^5 - 9a^6$
16. $12mn^2 - mn^2 = 11mn^2$
17. $10in^3 - 5in^3 = 5in^3$
18. $3\pi r^2 - 2\pi r^2 = \pi r^2$
19. $3xy^2 - 3x^2y = 3xy^2 - 3x^2y$
20. $15y^2 - 5y^2 - 2y^4 = 10y^2 - 2y^4$

ANSWERS: Worksheet 7

1. $x + x = 2x$
2. $2x^2 + 4x^2 = 6x^2$
3. $3xy^4 + 8xy = 3xy^4 + 8xy$
4. $x^3 + 2x^3 = 3x^3$
5. $ab^3 + ab^3 + 2ab^3 = 4ab^3$
6. $5 + 6m = 5 + 6m$
7. $9x^4y^4 - 7x^4y^4 = 2x^4y^4$
8. $rs^3 + rs^4 = rs^3 + rs^4$
9. $7abc^8 + 8abc^8 = 15abc^8$
10. $12x^5y^3 - 6x^5y^3 = 6x^5y^3$

11. $7x^3 + 4x^3 - 1x^3 = \mathbf{10x^3}$

12. $5xy^2 + 2xy^2 + 3a^2 - a^2 = \mathbf{7xy^2 + 2a^2}$

13. $12ab^4 - 6ab^4 + 2ab^4 = \mathbf{8ab^4}$

14. $5x^2 + 2ab^3 - ab^3 + 3x^2 = \mathbf{8x^2 + ab^3}$

15. $ax^5 + 3ax^5 - 2ax^5 = \mathbf{2ax^5}$

16. $3qrs^3 + 6qrs^3 + 4qrs^2 - qrs^2 = \mathbf{9qrs^3 + 3qrs^2}$

17. $12xy^2 - 11xy^3 + 10xy^3 - 3a^2 = \mathbf{12xy^2 - xy^3 - 3a^2}$

18. $15abc^2 - 2abc^2 + 4abc^2 - abc^2 = \mathbf{16abc^2}$

19. $4xyz^4 + 3xy^4z - x^4yz + xyz^4 = \mathbf{5xyz^4 + 3xy^4z - x^4yz}$

20. $28mn^{10} - 14mn^{10} + 2x^2y^4z^5 + 17x^2y^4z^5 - 3mn^{10} + x^3y^4z^6 = \mathbf{11mn^{10} + 19x^2y^4z^5 + x^3y^4z^6}$

ANSWERS: Worksheet 8

1. $x \cdot x^3 = x^4$
2. $xy \cdot xyz = x^2y^2z$
3. $a^4 \cdot ab \cdot b^3 = a^5b^4$
4. $2x \cdot 4x^2y = 8x^3y$
5. $3x^5y^2z^4 \cdot 7xyz = 21x^6y^3z^5$
6. $8ab \cdot 6ab \cdot ab = 48a^3b^3$
7. $7y \cdot 2x = 14xy$
8. $3a \cdot 4a = 12a^2$
9. $x^7 \cdot 2x^3 = 2x^{10}$
10. $4mn^{10} \cdot -6mn^2 = -24m^2n^{12}$

11. $-1/2a(3b) = -1.5ab$
12. $-3b(4x^2) = -12bx^2$
13. $\frac{5}{6}xy \cdot \frac{1}{3}y^2 = \frac{5}{18}xy^3$
14. $-5a^2 \times -6a^3 = 30a^5$
15. $-a(-a^4) = a^5$
16. $(a)^2 = a^2$
17. $3^2 \times 3^2 = 81$
18. $-b^2(-c^3) = b^2c^3$
19. $-\frac{1}{8}a\left(\frac{3}{4}c\right) = -\frac{3}{32}ac$
20. $a^3b^4c^3 \cdot abc^2 = a^4b^5c^5$

ANSWERS: Worksheet 9

1. $x^5 \div x^3 = x^2$

2. $12y^6 \div 2y^2 = 6y^4$

3. $\dfrac{a^7}{a^4} = a^3$

4. $\dfrac{9y^9}{3y^4} = 3y^5$

5. $\dfrac{2}{7}a^4 \div \dfrac{1}{3}a^3 = \dfrac{6}{7}a$

6. $\dfrac{36m^5n^4}{9m^2n} = 4m^3n^3$

7. $\dfrac{42x^4y^6z^3}{7x^2y^3z} = 6x^2y^3z^2$

8. $\dfrac{4ab^2}{2a} = 2b^2$

9. $\dfrac{-8b^2}{b^2} = -8$

10. $\dfrac{x^2y^2}{xy} = xy$

11. $\dfrac{\frac{2}{5}a^2}{\frac{1}{3}a} = \dfrac{2}{5}a^2 \div \dfrac{1}{3}a = \dfrac{6}{5}a$

12. $\dfrac{-9x^4y^2}{-3x^2y} = 3x^2y$

ANSWERS: Worksheet 10

1. $\dfrac{a^2}{a^4} = \dfrac{1}{a^2}$

2. $\dfrac{18y^3}{3y^6} = \dfrac{6}{y^3}$

3. $a^2 \div a^6 = \dfrac{1}{a^4}$ (Picture this one as $\dfrac{a^2}{a^6}$)

4. $48x^2 \div 8x^3 = \dfrac{6}{x}$

5. $\dfrac{49mn^4}{7m^2n^6} = \dfrac{7}{mn^2}$

6. $\dfrac{36y^6z}{6x^2y^3z^3} = \dfrac{6y^3}{x^2z^2}$

7. $42y^5 \div 6y^7 = \dfrac{7}{y^2}$

8. $\dfrac{24ab^2}{4a^3} = \dfrac{6b^2}{a^2}$

9. $\dfrac{-9cd^2}{3c^2d^5} = \dfrac{-3}{cd^3}$

10. $\dfrac{-16x^3y^2}{-4xy^6} = \dfrac{4x^2}{y^4}$

11. $\dfrac{\frac{3}{7}a^2b}{\frac{2}{3}ab^3} = \dfrac{9a}{14b^2}$ (divide the fractions)

12. $\dfrac{-32x^2y^2z^2}{4x^2y^7z} = \dfrac{-8z}{y^5}$

ANSWERS: Worksheet 11

1. Common factor = 2a

$$\frac{\cancel{2ab}+\cancel{4a}+\cancel{10ab^2}}{\cancel{2a}} =$$
(1b, 2, 5b²) / 1

$$\frac{b+2+5b^2}{1} = b+2+5b^2$$

2. Common factor = 5x

$$\frac{\cancel{10x}-\cancel{25xy}-\cancel{15x^3}}{\cancel{5x^2}} =$$
(2, -5y, -3x²) / 1x

$$\frac{2-5y-3x^2}{x}$$

3. Common factor = 7

$$\frac{\cancel{28mn}+\cancel{14m}-\cancel{49n^2}}{\cancel{7m^2n^2}} =$$
(4mn, 2m, -7n²) / 1

$$\frac{4mn+2m-7n^2}{m^2n^2}$$

4. Common factor = -6k

$$\frac{\cancel{42k}-\cancel{30k^2}+\cancel{18k^3}}{\cancel{-6k^3}} =$$
(-7, +5k, -3k²) / 1k²

$$\frac{-7+5k-3k^2}{k^2}$$

5. Common factor = -3y²

$$\frac{\cancel{-27y^4}-\cancel{9y^3}-\cancel{18y^2}}{\cancel{-3y^7}} =$$
(9y², 3y, 6) / y⁵

$$\frac{9y^2+3y+6}{y^5}$$

6. Common factor = 8

$$\frac{\cancel{-64b}-\cancel{48ab}+\cancel{56a^2}}{\cancel{8a}} =$$
(-8b, -6ab, 7a²) / 1a

$$\frac{-8b-6ab+7a^2}{a}$$

ANSWERS: Worksheet 11 (Continued)

7. Common factor = 4rs

$$\frac{16rs^2 + 12r^2s + 20r^3s^3t}{4rst^2} =$$

(cancellations shown: $4s$, $3r$, $5r^2s^2t$ over numerator; 1 under denominator)

$$\frac{4s + 3r + 5r^2s^2t}{t^2}$$

8. Common factor = w

$$\frac{11w^3 + 4w + 9w^2}{2w^3} =$$

(cancellations shown: $11w^2$, 4, $9w$ over numerator; $2w^2$ under denominator)

$$\frac{11w^2 + 4 + 9w}{2w^2}$$

9. Common factor = ab

$$\frac{ab - ab^2 - ab^3}{ab^2} =$$

(cancellations shown: 1, $-b$, $-b^2$ over numerator; b under denominator)

$$\frac{1 - b - b^2}{b}$$

10. Common factor = 11

$$\frac{11d + 22ef^2 + 121f}{11def} =$$

(cancellations shown: $1d$, $2ef^2$, $11f$ over numerator; $1def$ under denominator)

$$\frac{d + 2ef^2 + 11f}{def}$$

11. Common factor = $2xy^2$

$$\frac{6xy^2z + 4xy^2z + 8xy^2}{12xy^2z} =$$

(cancellations shown: $3z$, $2z$, $+4$ over numerator; $6z$ under denominator)

$$\frac{3z + 2z + 4}{6z} = \frac{5z + 4}{6z}$$

12. Common factor = $-9c$

$$\frac{-54c + 45c + 9c^2}{-9c^4} =$$

(cancellations shown: 6, -5, $-c$ over numerator; c^3 under denominator)

$$\frac{6 - 5 - c}{c^3} = \frac{1 - c}{c^3}$$

ANSWERS: Worksheet 12

1. $\dfrac{2+a}{2}$ = No factors

2. $\dfrac{\cancel{10x}}{\cancel{2x}} = \dfrac{5}{1} = 5$ The common factor is 2x.

3. $\dfrac{7\cancel{ab}^2}{\cancel{ab}} = \dfrac{7b}{1} = 7b$ The common factor is ab

4. $\dfrac{x+y+z^3}{xyz^5}$ = No common factors

5. $\dfrac{7+a}{a}$ = No factors

6. $\dfrac{\cancel{24a}}{\cancel{3a}} = \dfrac{8}{1} = 8$ The common factor is 3a.

7. $\dfrac{\cancel{5}ab+\cancel{5}b}{\cancel{5}} = \dfrac{ab+b}{1} = ab+b$ The common factor is 5.

8. $\dfrac{\cancel{3}(a+b)}{\cancel{3}} = \dfrac{a+b}{1} = a+b$ The common factor is 3.

9. $\dfrac{a+b}{a}$ = No factors

10. $\dfrac{4ac}{ac} = \dfrac{4}{1} = 4$ The common factor is ac.

11. $\dfrac{(y-1)}{y}$ = No factors.

ANSWERS: Worksheet 13

Rewrite each value without parentheses or without an exponent. The first one is done for you.

1. $(5x)^2 = 5^2 x^2 = 25x^2$
2. $(6+2)^2 = 8^2 = 64$
3. $(xy)^3 = x^3 y^3$
4. $(2ab)^4 = 2^4 a^4 b^4 = 16 a^4 b^4$
5. $(x+3)^2 = (x+3)(x+3)$
6. $(10-7)^3 = (10-7)(10-7)(10-7) = 3^3 = 27$
7. $(12-a)^2 = (12-a)(12-a)$
8. $(a+b)^3 = (a+b)(a+b)(a+b)$
9. $(3y)^2 = 3^2 y^2 = 9y^2$
10. $(4xy)^2 = 4^2 x^2 y^2 = 16 x^2 y^2$

11. $2x^4 + 4x^4 = 6x^4$
12. $6mn^2 - 3mn^2 = 3mn^2$
13. $5xy^3 \cdot -3x^2 y^2 = -15 x^3 y^5$
14. $\dfrac{24 a^2 b^2}{3ab} = \dfrac{8ab}{1} = 8ab$
15. $8x + 8x^2 = 8x + 8x^2$ ⎫
16. $34a^3 - 3b^3 = 34a^3 - 3b^3$ ⎬ → Not like terms.
17. $9ab^4 c^2 \cdot 8a^3 b^3 c = 72 a^4 b^7 c^3$
18. $\dfrac{4xy^2}{2xy^4} = \dfrac{2}{y^2}$
19. $3xyz^3 + 4xyz^2 = 3xyz^3 + 4xyz^2$
20. $4r^2 s^2 t^3 \cdot 11 rs^3 t^2 \cdot 2rst = 88 r^4 s^6 t^6$

ANSWERS: Worksheet 14

1. $\sqrt{16} = 4$
2. $\sqrt{4} = 2$
3. $\sqrt{49} = 7$
4. $\sqrt{81} = 9$
5. $\sqrt{25} = 5$
6. $\sqrt{9} = 3$
7. $\sqrt{100} = 10$
8. $\sqrt{144} = 12$
9. $\sqrt{121} = 11$
10. $\sqrt{36} = 6$

ANSWERS: Worksheet 15

1. $\sqrt[3]{8} = 2$
2. $\sqrt[3]{64} = 4$
3. $\sqrt[3]{216} = 6$
4. $\sqrt[3]{27} = 3$
5. $\sqrt[3]{125} = 5$
6. $\sqrt[3]{-64} = -4$
7. $\sqrt[3]{-8} = -2$
8. $\sqrt[3]{-27} = -3$
9. $\sqrt[3]{-1000} = -10$
10. Bonus problem: $\sqrt[4]{81} = 3$
11. $\sqrt[3]{8} + \sqrt[3]{8} = 4$
12. $\sqrt[3]{8} \cdot \sqrt[3]{-8} = -4$
13. $\dfrac{\sqrt[3]{-216}}{\sqrt[3]{8}} = -3$

ANSWERS: Worksheet 16

1. $\sqrt{\dfrac{9}{36}} = \dfrac{3}{6}$

2. $\sqrt{\dfrac{2}{25}} = \dfrac{\sqrt{2}}{5}$

3. $\sqrt{\dfrac{4}{x}} = \dfrac{2}{\sqrt{x}}$

4. $\sqrt{\dfrac{7}{64}} = \dfrac{\sqrt{7}}{8}$

5. $\sqrt{\dfrac{16}{81}} = \dfrac{4}{9}$

6. $\sqrt{\dfrac{b}{121}} = \dfrac{\sqrt{b}}{11}$

7. $\sqrt{\dfrac{a}{c}} = \dfrac{\sqrt{a}}{\sqrt{c}}$

8. $\sqrt{\dfrac{7}{9}} = \dfrac{\sqrt{7}}{3}$

9. $\sqrt{\dfrac{9}{10}} = \dfrac{3}{\sqrt{10}}$

10. $\sqrt{\dfrac{25}{49}} = \dfrac{5}{7}$

ANSWERS: Chapter 2 Review Test

1. $a^3 + a^3 = \mathbf{2a^3}$
2. $5ac^2 + ac^2 = \mathbf{6ac^2}$
3. $7y^2 + 5y^2 + 2y^2 = \mathbf{14y^2}$
4. $9x^4y^5z^6 + 7x^4y^5z^6 = \mathbf{16x^4y^5z^6}$
5. $4b^3 + 2b^3 + 6y^2 = \mathbf{6b^3 + 6y^2}$
6. $6ab^3 + 4xy^3 + 4ab^3 + 6xy^3 = \mathbf{10ab^3 + 10xy^3}$

7. $12m^4 - 7m^4 = \mathbf{5m^4}$
8. $-13y^3 - 8y^3 = \mathbf{-21y^3}$
9. $5r^2 - (-r^2) = \mathbf{6r^2}$
10. $8^2 - 6^2 = 64 - 36 = \mathbf{28}$
11. $6jk^2 - m^3 - jk^2 = \mathbf{5jk^2 - m^3}$
12. $-abc^4 - (abc^4) = \mathbf{-2abc^4}$

13. $9a^2 \cdot 3a^2 = \mathbf{27a^4}$
14. $xy^3 \cdot xy = \mathbf{x^2y^4}$
15. $3^2 \cdot x^2 = \mathbf{9x^2}$
16. $(7e^3f^2g)^2 = \mathbf{49e^6f^4g^2}$
17. $-\frac{2}{5}x^2 \cdot \frac{2}{3}x^3 = \mathbf{-\frac{4}{15}x^5}$
18. $(-2xy)^3 = \mathbf{-8x^3y^3}$

19. $\frac{a^5}{a^2} = \frac{a^3}{1} = \mathbf{a^3}$
20. $\frac{a^3}{a^5} = \mathbf{\frac{1}{a^2}}$

21. $\frac{24x^2y^3}{4xy} = \frac{6xy^2}{1} = \mathbf{6xy^2}$
22. $\frac{-48m^6}{8m^{12}} = \mathbf{\frac{-6}{m^6}}$

23. $54ab^2 \div -6ab^4 = \mathbf{-\frac{9}{b^2}}$
24. $\frac{10ab^3 + 8a^2 + 6b}{2ab} = \mathbf{\frac{5ab^3 + 4a^2 + 3b}{ab}}$

25. $\frac{32c^2 + 8a^2c^2 - 12abc^2}{4ac^2} = \mathbf{\frac{8 + 2a^2 - 3ab}{a}}$
26. $\frac{-a+b-a}{ab^2} =$ There are no common Factors.

Fill in the blank.

27. Canceling **F**actors is **F**ine, but canceling **T**erms is **T**rouble.

28. $\frac{a+b}{ab}$
No factors to cancel.

29. $\frac{5ab}{7ab^2} = \mathbf{\frac{5}{7b}}$

30. $\frac{6 + a^3}{6a^3}$
No factors to cancel.

ANSWERS: Chapter 2 Test Continued...

31. $\dfrac{18a^2b^2c^2}{6abc} = \dfrac{3abc}{1} = \mathbf{3abc}$ 32. $\dfrac{7(a+b)}{7xy} = \dfrac{a+b}{xy}$ 33. $\dfrac{45y + 18y^2 + 27yz}{-9yz^2} = \dfrac{-5 - 2y - 3z}{z^2}$

34. Which one of the following terms is the only possible answer to the volume of a cube?

        8in²   9xy²   (64in³)   1/2in⁴   7ft²

35. Which one of the following terms is the only possible answer to the area of a square?

        5in³   4ft.   (9cm²)   8 miles   π

36. $\sqrt{81} = \pm 9$ 37. $\sqrt[3]{-125} = -5$ 38. $\sqrt{\dfrac{16}{25}} = \dfrac{4}{5}$

39. $\sqrt[2]{25} = \pm 5$ 40. $\sqrt{\dfrac{9}{b}} = \dfrac{3}{\sqrt{b}}$ 41. $\sqrt{49} - \sqrt[3]{-8} =$
                                                                                                                              $7 - -2 = 9$

42. $(\sqrt[3]{216})^2 =$ 43. $\sqrt{-9} =$ 44. $\dfrac{\sqrt[3]{-512}}{\sqrt{16}} = \dfrac{-8}{4} = -2$

    $(6)^2 = 36$ **Not a real number.**

45. $\dfrac{a^2 - 4ac}{4a^2} = \dfrac{a - 4c}{4a}$ 46. $\dfrac{28a^3 + \sqrt{25} - 3a^3}{5a^3} =$ 47. $\dfrac{-3x^2 - 5x^3 + 9x}{-x^4} =$

                                                                               $\dfrac{28a^3 + 5 - 3a^3}{5a^3} =$ $\dfrac{3x + 5x^2 - 9}{x^3}$

                                                                               $\dfrac{28a^3 - 3a^3 + 5}{5a^3} =$

                                                                               $\dfrac{25a^3 + 5}{5a^3} =$

                                  $\dfrac{5a^3 + 1}{a^3}$ (I factored out the 5)

48. $33x^2y^3 - 14a^2b + x^2y^3 - 9x^2 - 3a^2b + 8x^2 - (4^2x^2y^3) = \mathbf{18x^2y^3 - 17a^2b - x^2}$
    $33x^2y^3 + x^2y^3 - (16x^2y^3) = 18x^2y^3$ $-14a^2b - 3a^2b = -17a^2b$ $-9x^2 + 8x^2 = -x^2$

ANSWERS: Worksheet 17

1. **a + b = b + a** is an example of which law?
   COMMUTATIVE LAW OF ADDITION

2. **$2a^2 + 5a^2 = 5a^2 + 2a^2$** is an example of which law?
   COMMUTATIVE LAW OF ADDITION

3. What does the root word "commute" mean?
   TO EXCHANGE

4. What does the root word "associate" mean?
   A MEMBER OF A GROUP

5. 5 + (3 + 8) = (5 + 3) + 8 is an example of which law?
   ASSOCIATIVE LAW OF ADDITION

6. 2x + (5x + b) = (2x + 5x) + b is an example of which law?
   ASSOCIATIVE LAW OF ADDITION

7. Use the Associative Law of Addition to rewrite the following expression.
   (19 + a) + b = 19 + (a + b)

8. Use the Commutative law of Addition to rewrite the following.
   7a + 2 + 3a = 7a + 3a + 2

ANSWERS: Chapter 3 Review Test

1.   a. .5   14   27   .75   0 **Neither**   c. 5   7   110   1   3   287 **Natural**
       b. -2   5   25   -243   0   17 **Integers**   d. -9   -136   -298   -600   ½ **Neither**

Simplify the following.
2. $2y^2z^3 + 5y^2z^3$ = **$7y^2z^3$**
3. $9xyz \cdot 7x^2y^3z$ = **$63x^3y^4z^2$**
4. $\dfrac{14d^3e^2f}{7def}$ = **$2d^2e$**

Solve the following problems.
5. |-9| + |9| = **18**   6. |13| - |-5| = **8**   7. |1/2| x |-25| = **12.5**
8.   -27 + -7 = **-34**   9.   -5 x 8 = **-40**   10.   17 - -5 = **22**
11.   -56 ÷ -7 = **8**   12.   -45.9 - -13.8 = **-32.1**   13.   -90 + 45 = **-45**

Rewrite each expression using the Associative Law of Addition. The first one is done for you.

*.   (3a + a) + 5a = (3a + 5a) + a
14.   7xy + (xy + 2) = **(7xy + xy) + 2**
15.   5 + (15 + 8) = **(5 + 15) + 8**

Rewrite each expression using the Associative Law of Multiplication. The first one is done for you.

*.   6(2a · 3) = 2a(6 · 3)
16.   (7b · 3) · 3b = **3(7b · 3b)**
17.   x · (2x · ½) = **½ (2x · x)**

Solve the following.
18. $\sqrt{\dfrac{16}{49}} = \dfrac{4}{7}$ 19. $\sqrt{\dfrac{b}{64}} = \dfrac{\sqrt{b}}{8}$ 20. $\dfrac{\frac{5}{7}}{\frac{3}{4}} = \dfrac{5}{7} \times \dfrac{4}{3} = \dfrac{20}{21}$ 21. $\dfrac{12a^4 - 6a}{3a^2} = \dfrac{4a^3 - 2}{a}$

ANSWERS: Worksheet 19

1. $4(5 + 7) =$
   $(4 \cdot 5) + (4 \cdot 7) =$
   $20 + 28 = 48$

2. $7(8 + a) =$
   $(7 \cdot 8) + (7 \cdot a) = 2$
   $56 + 7a$

3. $8a(5a + 2) =$
   $(8a \cdot 5a) + (8a \cdot 2) =$
   $40a^2 + 16a$

4. $3x^2(2x + 9x^3) =$
   $(3x^2 \cdot 2x) + (3x^2 \cdot 9x^3)$
   $6x^3 + 27x^5$

5. $2xy(2xy + 5) =$
   $(2xy \cdot 2xy) + (2xy \cdot 5) =$
   $4x^2y^2 + 10xy$

6. $2x^2y^2(3x + 4y) =$
   $(2x^2y^2 \cdot 3x) + (2x^2y^2 \cdot 4y) =$
   $6x^3y^2 + 8x^2y^3$

7. $-7(5a + 8x) =$
   $(-7 \cdot 5a) + (-7 \cdot 8x) =$
   $-35a + -56x$

8. $-10(a + b) =$
   $-10a + -10b$

9. $6a^3b^2(-2ab + 3a^3b^2) =$
   $(6a^3b^2 \cdot -2ab) + (6a^3b^2 \cdot 3a^3b^2) =$
   $-12a^4b^3 + 18a^6b^4$

ANSWERS: Worksheet 19 page 2

10. $-\frac{1}{2}a\left(-\frac{2}{3}a + \frac{3}{4}b\right) =$
    $\left(-\frac{1}{2}a \cdot -\frac{2}{3}a\right) + \left(-\frac{1}{2}a \cdot \frac{3}{4}b\right) =$
    $\frac{2}{6}a^2 - \frac{3}{8}ab$

11. $\frac{3}{8}b^2\left(-\frac{1}{4}b - \sqrt{\frac{4}{16}}\right) =$
    $\left(\frac{3}{8}b^2 \cdot -\frac{1}{4}b\right) - \left(\frac{3}{8}b^2 \cdot \frac{\overset{1}{2}}{\underset{4}{4}}\right) =$

    $-\frac{3}{32}b^3 - \frac{3}{16}b^2$

ANSWERS: Worksheet 20

Use the Distributive Property of Multiplication to simplify the following.

1.   -1(3y + 2x) $-3y - 2x$

2.   -(5m - 7n) $-5m + 7n$

3.   -(-7a - 9b) $7a + 9b$

4.   -1(3x + -5y) $-3x + 5y$

5.   -2(a + b) $-2a - 2b$

6.   -5a(5a - 2a) $-25a^2 + 10a^2 = -15a^2$

7.   -x(x² - -y) $-x^3 - xy$

8.   3a(-a + -b) $-3a^2 - 3ab$

9.   7(9 - 3) $63 - 21 = 42$

10.   -10a(-3a + 7a²) $30a^2 - 70a^3$

11. $-\frac{3}{5}mn\left(-\frac{5}{6} + mn\right) =$ $\left(-\frac{3}{5}mn \cdot -\frac{5}{6}\right) + \left(-\frac{3}{5}mn \cdot mn\right) = \frac{15}{30}mn - \frac{3}{5}m^2n^2$

12. $-\sqrt[3]{64}(-\sqrt[3]{8} + -\sqrt[3]{27})$ $-4(-2 - 3) =$ $8 + 12 = 20$

ANSWERS: Worksheet 21

1. $6x(-4y + 8x - 7y)$
   $(6x \cdot -4y) + (6x \cdot 8x) - (6x \cdot 7y)$
   $(-24xy) + (48x^2) - (42xy)$
   $48x^2 - 66xy$

2. $3a(4 - 6a + 2a^2)$
   $(3a \cdot 4) - (3a \cdot 6a) + (3a \cdot 2a^2)$
   $(12a) - (18a^2) + (6a^3)$
   $12a - 18a^2 + 6a^3$

3. $-(-x + 4y - 3)$
   Just change the signs.
   $x - 4y + 3$

4. $-6xy(5y + 2xy - 4y^2)$
   $(-6xy \cdot 5y) + (-6xy \cdot 2xy) - (-6xy \cdot 4y^2)$
   $-30xy^2 - 12x^2y^2 + 24xy^3)$

5. $4(6 + 2x - 3x^2 - 6y)$
   $(4 \times 6) + (4 \cdot 2x) - (4 \cdot 3x^2) - (4 \times 6y)$
   $(24) + (8x) - (12x^2) - (24y)$
   $24 + 8x - 12x^2 - 24y$

6. $-3xy^3(-4x^2y^2 + 2y - 6x)$
   $(-3xy^3 \cdot -4x^2y^2) + (-3xy^3 \cdot 2y) - (-3xy^3 \cdot 6x)$
   $(12x^3y^5) + (-6xy^4) - (-18x^2y^3)$
   $12x^3y^5 - 6xy^4 + 18x^2y^3$

7. $7a(8a + b + 2c)$
   $7a(8a) + 7a(b) + 7a(2c)$
   $56a^2 + 7ab + 14ac$

8. $a^2(4x - a - 7x^3)$
   $a^2(4x) - a^2(a) - a^2(7x^3)$
   $4a^2x - a^3 - 7a^2x^3$

ANSWERS: Chapter 4 Review Test

Simplify the following, by using the Distributive Law of Multiplication.

1. $5a(a+b)$
   $5a^2 + 5ab$

2. $2x^2(3x+5)$
   $6x^3 + 10x^2$

3. $-\frac{3}{8}a\left(\frac{3}{4}a + \frac{4}{7}b\right)$
   $-\frac{9}{32}a^2 - \frac{12}{56}ab$

4. $-(3a + 2a - 6b)$
   $-3a - 2a + 6b = -5a + 6b$

5. $\quad 4y + 3a^2 + 10y$
   $\underline{-2y - a^2\ \ - 2y}$
   $2y + 2a^2 + 8y = 10y + 2a^2$

6. $3a(6b^2 - 4ab + 8)$
   $18ab^2 - 12a^2b + 24a$

7. $5b(-3b - 7b + 3b^2)$
   $-15b^2 - 35b^2 + 15b^3 = -50b^2 + 15b^3$

8. $-3x^2y^2(-8y^5 + 9x^5 - 6y^3)$
   $24x^2y^7 - 27x^7y^2 + 18x^2y^5$

ANSWERS: Worksheet 22

Find all the factors of the following numbers.

1. 30   2, 3, 5, 6, 10, 15

2. 12   2, 3, 4, 6

3. 55   5, 11

4. 54   2, 3, 6, 9, 18, 27

5. 24a   a, 2, 3, 4, 6, 8, 12,

6. 9b   3, 3, b

7. x^2 x, x

8. 18y   2, 3, 6, 9, y

9. $7ab^2$ 7, a, b, b

10. 64xyz   2, 4, 8, 16, 32, x, y, z

ANSWERS: Worksheet 23

List all the prime numbers between 2-50.

2, 3, 5, 7, 11, 13, 17, 19, 23, 29, 31, 37, 41, 43, 47

Factor the following terms.

1.	$3a$	3 a
2.	$2y^2$	2 y y
3.	$11(a + b)$	11 (a + b)
4.	$3ab^3c$	3 a b b b c
5.	$7(8 - n)$	7 (8 - n)
6.	$5xyz^2$	5 x y z z
7.	$a(b)$	a b
8.	$41a^3b^3c$	41 a a a b b b c
9.	pqr	p q r
10.	10^2	10 10
11.	$\frac{1}{2}bh$	½ b h
12.	bhw	b h w
13.	$d\pi$	d π
14.	πr^2	π r r

ANSWERS: Worksheet 24

1. 18
   / \
   6   3
  /|\ |
  2 3 3

2. 36
   / \
   6   6
  /\ /\
  2 3 2 3

3. 49
   / \
   7   7

4. 100
   / \
  10   10
  /\   /\
  2 5 2 5

5. 62
   /\
  31 2

6. 63
   / \
   9   7
  /\ |
  3 3 7

7. 77
   /\
  7 11

8. 32
   / \
   8   4
  /\   /\
  2 4 2 2
  /\ | |
  2 2 2 2 2

9. 47
   /
   47 is already prime

10. 82
    /\
   41 2

11. 52
    /\
   26 2
   /\ |
  13 2 2

12. 81
    /\
    9 9
   /| /\
   3 3 3 3

13. 48
    /\
    6 8
   /\ /|\
   2 3 2 4

   2 3 2 2 2

14. 64
    /\
    8 8
   /| /|\
   2 4 2 4

   2 2 2 2 2 2

15. 65
    /\
    5 13

229

ANSWERS: Worksheet 25

1. $5a + 10a$ $5a(1 + 2)$

2. $7x + 21x$ $7x(1 + 3)$

3. $x^2 + x^3$ $x^2(1 + x)$

4. $3xy - 6x$ $3x(y - 2)$

5. $4ab - 2b$ $2b(2a - 1)$

6. $12xy^2 - 4y$ $4y(3xy - 1)$

7. $3abc^3 - 9ab$ $3ab(c^3 - 3)$

Next, look at each one of your answers from problems 1 – 7 and distribute each one below. The example from above has been done for you.

* $2a(3 + 1)$ $= (2a \times 3) + (2a \times 1) =$ $6a + 2a$

8. $5a(1 + 2)$
    $(5a \times 1) + (5a \times 2)$
    $5a + 10a$

9. $7x(1 + 3)$
    $7x(1) + 7x(3)$
    $7x + 21x$

10. $x^2(1 + x)$
    $(x^2 \cdot 1) + (x^2 \cdot x)$
    $x^2 + x^3$

11. $3x(y - 2)$
    $(3x \cdot y) - (3x \cdot 2)$
    $3xy - 6x$

12. $2b(2a - 1)$
    $2b(2a) - 2b(1)$
    $4ab - 2b$

13. $4y(3xy - 1)$
    $12xy^2 - 4y$

14. $3ab(c^3 - 3)$
    $3abc^3 - 9ab$

ANSWERS: Chapter 5 Review Test

1.
   a. 555, 678, -945, -122   Integers
   b. 1, 2, 0, 4, 22, 11   Integers
   c. 956, 33, 25, 1,094   Natural Numbers
   d. $\frac{1}{4}$ $\frac{1}{2}$ $\frac{3}{4}$ $\frac{1}{1}$ Neither

Simplify the following.

2. $17xyz^3 - 12xyz^3 = 5xyz^3$
3. $a^3b^2 + 2a^2b^2 = a^3b^2 + 2a^2b^2$
4. $6m^4n^3 \cdot 8mn^2 = 48m^5n^5$

Give the absolute value for each number.

5. $|-8| = 8$
6. $|23| = 23$
7. $|-\pi| = 3.14$
8. $|-d| = d$

Do the math.

9. $-7 \cdot -8 = 56$
10. $-23 + -7 = -30$
11. $34 - -8 = 42$
12. $-11 \cdot 16 = -176$
13. $-3/4 \cdot 5/8 = -15/32$
14. $-m \cdot -n = mn$
15. $2a + -5a = -3a$
16. $-3a(7a) = -21a^2$

ANSWERS: Chapter 5 Review Test page 2

The following describes which law of addition?

17. $a + b = b + a$ — Commutative
18. $3a + 4a = 4a + 3a$ — Commutative
19. $2 + (3 + 4a) = (2 + 3) + 4a$ — Associative
20. $8x^3y^2 + 3a + 5x^3y^2 = 5x^3y^2 + 8x^3y^2 + 3a$ — Commutative

The following describes which law of multiplication?

21. $2a \cdot 3a^2 = 3a^2 \cdot 2a$ — Commutative
22. $3a(2a \cdot 4x) = 4x(3a \cdot 2a)$ — Associative
23. $5 \times 7 \times 10 = 7 \times 10 \times 5$ — Commutative
24. $6x(3x - 2a) = (6x \cdot 3x) - (6x \cdot 2a)$ — Distributive

Solve the following problems.

25. $2(3 + 5)$
     $(2 \times 3) + (2 \times 5)$
     $6 + 10 = 16$

26. $2a(4a + 6a) =$
     $(2a \times 4a) + (2a \times 6a)$
     $8a^2 + 12a^2$
     $20a^2$

27. $2x(5x - 3x) =$
     $(2x \cdot 5x) - (2x \cdot 3x)$
     $10x^2 - 6x^2$
     $4x^2$

28. $4m(3m - 2m) =$
     $(4m \cdot 3m) - (4m \cdot 2m)$
     $12m^2 - 8m^2$
     $4m^2$

ANSWERS: Chapter 5 Review Test page 3

29. $-(3y + 8x - 7y) =$
    $(-1 \cdot 3y) + (-1 \cdot 8x) - (-1 \cdot 7y)$
    $(-3y) + (-8x) - (-7y)$
    $-3y - 8x + 7y$
    (Or just change the signs)
    $-3y - 8x + 7y$
    $-8x + 4y$

30. $-8x^2y(3x - 8xy^3) =$
    $(-8x^2y \cdot 3x) - (-8x^2y \cdot 8xy^3)$
    $(-24x^3y) - (-64x^3y^4)$
    $-24x^3y + 64x^3y^4$

Factor the following terms down to all prime factors.

31. $32x^2y$ 2 2 2 2 2 x x y
32. $25y^2$ 5 5 y y
33. 42   2 3 7
34. 72ab   3 3 2 2 2 a b
35. $112x^3$ 2 2 2 2 7 x x x

Factor out the common factor

36. $3\pi r^3 + 27r^3$ $3r^3(\pi + 9)$
37. $16n^2 - 4n^3$ $4n^2(4 - n)$
38. $49a^2b + 49ab^2$ $7ab(7a + 7b)$
39. $-3x^2 + 27x^2$ $3x^2(-1 + 9)$
40. $\sqrt{7}ab + \sqrt{7}xy$ $\sqrt{7}(ab + xy)$

ANSWERS: Chapter 5 Review Test continued

Use the Distributive Law of Multiplication, to solve the following.

41. $4xy(5y + 3x + 2)$
    $20xy^2 + 12x^2y + 8xy$

42. $-2a(9ab - 5b + a^2)$
    $-18a^2b + 10ab - 2a^3$

43. $-\frac{5}{9}\left(ab^2 - \frac{4}{3}a^3\right)$
    $-\frac{5}{9}ab^2 + \frac{20}{27}a^3$

Divide the following terms.

44. $\frac{3x^2 + 6xy - 9x^3}{x^2} = \frac{3x + 6y - 9x^2}{x}$ factor out the 3 $= \frac{3(x + 2y - 3x^2)}{x}$

45. $\frac{-3abc}{-12ab^2c^3} = \frac{1}{4bc^2}$

46. $\frac{14y + 21xyz - 56y^3z^2}{7y^3} = \frac{2 + 3xz - 8y^2z^2}{y^2}$

ANSWERS: Worksheet 26

1. $4r(6r - 9x + 5t)$
   $(4r \cdot 6r) - (4r \cdot 9x) + (4r \cdot 5t)$
   $(24r^2) - (36rx) + (20rt)$
   $24r^2 - 36rx + 20rt$

2. $7x(8y + 3x - 2x^2)$
   $(56xy) + (21x^2) - (14x^3)$
   $56xy + 21x^2 - 14x^3$

3. $-5y(3y + 2x - 4y)$
   $-5y \cdot +3y \quad -5y \cdot +2x \quad -5y \cdot -4y$
   $-15y^2 \;\; -10xy \;\; +20y^2$

4. $-(9x + 10x^2 - 5y)$
   Just change the signs
   $-9x \;\; -10x^2 + 5y$

5. $3x^2y(2x + 4y - 1)$
   $(3x^2y \cdot 2x) + (3x^2y \cdot 4y) - (3x^2y \cdot 1)$
   $(6x^3y) + (12x^2y^2) - (3x^2y)$
   $6x^3y + 12x^2y^2 - 3x^2y$

6. $2s^2(3s^2 - 5s + 6)$
   $(2s^2 \cdot 3s^2) - (2s^2 \cdot 5s) + (2s^2 \cdot 6)$
   $(6s^4) - (10s^3) + (12s^2)$
   $6s^4 - 10s^3 + 12s^2$

7. $5a^2(5a - 3a^2 + 4)$
   $(5a^2 \cdot 5a) - (5a^2 \cdot 3a^2) + (5a^2 \cdot 4)$
   $25a^3 - 15a^4 + 20a^2$

ANSWERS: Worksheet 26 continued

8. $-4a(a^2 - 2ab - b)$
   $(-4a \cdot a^2) - (-4a \cdot 2ab) - (-4a \cdot b)$
   $(-4a^3) - (-8a^2b) - (-4ab)$
   $-4a^3 + 8a^2b + 4ab$

9. $(-3xy)(x^2 - 2xy + y^2)$
   $(-3xy \cdot x^2) - (-3xy \cdot 2xy) + (-3xy \cdot y^2)$
   $(-3x^3y) - (-6x^2y^2) + (-3xy^3)$
   $-3x^3y + 6x^2y^2 - 3xy^3$

10. $(6m^2 - 2n^2 - 4mn)2m^3$
    $(2m^3 \cdot 6m^2) - (2m^3 \cdot 2n^2) - (2m^3 \cdot 4mn)$
    $(12m^5) - (4m^3n^2) - (8m^4n)$
    $12m^5 - 4m^3n^2 - 8m^4n$

ANSWERS: Worksheet 27

Try some on your own. Use the FOIL method to simplify the following.

1. $(3x - 2)(4x - 1)$
   $12x^2 - 3x - 8x + 2$
   $12x^2 - 11x + 2$

2. $(5x - 3)^2$
   $(5x - 3)(5x - 3)$
   $25x^2 - 15x - 15x + 9$
   $25x^2 - 30x + 9$

3. $(x + y)(x - y)$
   $x^2 - xy + xy - y^2$
   $x^2 - y^2$

4. $(5y + 2)(5y - 2)$
   $25y^2 - 10y + 10y - 4$
   $25y^2 - 4$

5. $(a + b)(a - b)$
   $a^2 - ab + ab - b^2$
   $a^2 - b^2$

6. $(2a + 7)(a - 3)$
   $2a^2 - 6a + 7a - 21$
   $2a^2 + a - 21$

7. $(3x - 2y)^2$
   $(3x - 2y)(3x - 2y)$
   $9x^2 - 6xy - 6xy + 4y^2$
   $9x^2 - 12xy + 4y^2$

ANSWERS: Worksheet 27 page 2

8. $(3m - 1)(2m - 5)$
   $6m^2 - 15m - 2m + 5$
   $6m^2 - 17m + 5$

9. $(2a + b)^2$
   $(2a + b)(2a + b)$
   $4a^2 + 2ab + 2ab + b^2$
   $4a^2 + 4ab + b^2$

10. $(m + n)^2$
    $m^2 + mn + mn + n^2$
    $m^2 + 2mn + n^2$

ANSWERS: Worksheet 28

Simplify the following polynomials.

1. $3a(5a + 6x - 7)$
   $15a^2 + 18ax - 21a$

2. $(2 + 3x)(2x - 8y)$
   $4x - 16y + 6x^2 - 24xy$

3. $-7m(m + n + 2x)$
   $-7m^2 - 7mn - 14mx$

4. $(4x + 5)(3x - 1)$
   $12x^2 - 4x + 15x - 5$
   $12x^2 + 11x - 5$

5. $(3xy^2 + 4y)(4x + 3xy - 2x)$
   $12x^2y^2 + 9x^2y^3 - 6x^2y^2 + 16xy + 12xy^2 - 8xy$
   $6x^2y^2 + 9x^2y^3 + 8xy + 12xy^2$

6. $(x - 5)^2$
   $(x - 5)(x - 5)$
   $x^2 - 5x - 5x + 25$
   $x^2 - 10x + 25$

7. $(7y + 3)^3$
   $(7y + 3)(7y + 3)(7y + 3)$
   $(49y^2 + 21y + 21y + 9)(7y + 3)$
   Switch them around and combine like terms.
   $(7y + 3)(49y^2 + 42y + 9)$
   $343y^3 + 294y^2 + 63y + 147y^2 + 126y + 27$
   Combine like terms.
   $343y^3 + 441y^2 + 189y + 27$

ANSWERS: Worksheet 28 page 2

8. $(5xyz + 3z - 8y)(6z - 9x + 5xyz)$
    $30xyz^2 - 45x^2yz + 25x^2y^2z^2 + 18z^2 - 27xz + 15xyz^2 - 48yz + 72xy - 40xy^2z$
    Combine like terms.
    $45xyz^2 - 45x^2yz + 25x^2y^2z^2 + 18z^2 - 27xz - 48yz + 72xy - 40xy^2z$

9. $-(7x^2y^3z^4 - 12xy^3 - 5x^2y^3z^4)$
    Just change the signs.
    $-7x^2y^3z^4 + 12xy^3 + 5x^2y^3z^4$
    Combine like terms.
    $-2x^2y^3z^4 + 12xy^3$

10. $(x - y^2)(y - x^2)(xy + xy)$
    $(x - y^2)(y - x^2)(xy + xy)$
    $(xy - x^3 - y^3 + x^2y^2)(xy + xy)$
    Switch them around, if you prefer.
    $(xy + xy)(xy - x^3 - y^3 + x^2y^2)$
    $x^2y^2 - x^4y - xy^4 + x^3y^3 + x^2y^2 - x^4y - xy^4 + x^3y^3$
    Combine like terms.
    $2x^2y^2 - 2x^4y - 2xy^4 + 2x^3y^3$

ANSWERS: Worksheet 29

Factor the polynomials.

1. $x^2 + 8x + 16$
   $(x + 4)(x + 4)$

2. $m^2 + 5m + 6$
   $(m + 3)(m + 2)$

3. $x^2 - 8x + 7$
   $(x - 1)(x - 7)$

4. $y^2 + 4y - 32$
   $(y + 8)(y - 4)$

5. $x^2 + 13x + 42$
   $(x + 7)(x + 6)$

6. $n^2 + 5n - 36$
   $(n - 4)(n + 9)$

7. $b^2 + 7b - 30$
   $(b + 10)(b - 3)$

8. $c^2 - 10c + 16$
   $(c - 8)(c - 2)$

9. $a^2 + 8a - 33$
   $(a + 11)(a - 3)$

10. $x^2 + 16x + 48$
    $(x + 12)(x + 4)$

ANSWERS: Chapter 6 Review Test

1. $-x^2 \cdot -x^3 = x^5$
2. $(-8) - (-5) = -3$
3. $4xy(-3xy) = -12x^2y^2$
4. $-3x^3(7x) = -21x^4$
5. $-5(5x^4) = -25x^4$
6. $-4x^2y^2 \cdot -4x^2y^2 = 16x^4y^4$
7. $-mn(m - n) = -m^2n + mn^2$

Name the law that is being used to rearrange each problem.

8. $4a + 3 + 2a = 2a + 4a + 3$ Commutative Law of Addition
9. $8a + 7 + 9 + 6a = (8a + 6a) + (7 + 9)$ Associative Law of Addition
11. $7xy(3x \cdot 2xy) = 3x(7xy \cdot 2xy)$ Associative Law of Multiplication
11. $(x)(5)(2) = 5 \cdot 2 \cdot x$ Commutative Law of Multiplication
12. $4a(3x \cdot 9a) = (4a \cdot 3x)(4a \cdot 9a)$ Distributive Law of Multiplication

Simplify the following using the Distributive Law of Multiplication.

13. $2a(5 - 3a)$
    $2a(5) - 2a(3a)$
    $10a - 6a^2$

14. $-5b(2a - 3b)$
    $-5b(2a) - -5b(3b)$
    $-10ab + 15b^2$

15. $m(2m + 4 + 6m^2)$
    $m(2m) + m(4) + m(6m^2)$
    $2m^2 + 4m + 6m^3$

ANSWERS: Chapter 6 Review Test page 2

16. $-5x(3x + 4y - 4x + 3)$
    $-5x(3x) + -5x(4y) - -5x(4x) + -5x(3)$
    $-15x^2 - 20xy + 20x^2 - 15x$
    $5x^2 - 20xy - 15x$

17. $(b + c - d)a$
    $(ab) + (ac) - (ad)$

Use the FOIL method to multiply the following.

18. $(b + 3)(b - 4)$
    $b^2 + 3b - 4b - 12$
    $b^2 - b - 12$

19. $(x - 4)(x - 5)$
    $x^2 - 4x - 5x + 20$
    $x^2 - 9x + 20$

20. $(y + 8)(y + 7)$
    $y^2 + 8y + 7y + 56$
    $y^2 + 15y + 56$

21. $(r - 9)(r + 2)$
    $r^2 + 2r - 9r - 18$
    $r^2 - 7r - 18$

22. $(m - 1)(m - 8)$
    $m^2 - 8m - 1m + 8$
    $m^2 - 9m + 8$

ANSWERS: Chapter 6 Review Test page 3

Factor the following polynomials.

23. $x^2 + 15x + 56$
    $(x + 8)(x + 7)$

24. $x^2 + 5x - 24$
    $(x + 8)(x - 3)$

25. $x^2 - 15x + 50$
    $(x - 10)(x - 5)$

26. $y^2 + 14y + 24$
    $(y + 12)(y + 2)$

27. $x^2 - 12x + 20$
    $(x - 10)(x - 2)$

28. $m^2 + 2m - 48$
    $(m + 8)(m - 6)$

29. $x^2 + 5x - 36$
    $(x + 9)(x - 4)$

30. $x^2 - 12x - 13$
    $(x - 13)(x + 1)$

ANSWERS: Worksheet 30

Rearrange the equations, to fit into $ax^2 + bx + c = 0$. Write down which term is a, which is b, and which is c. The first one has been done for you.

* $-8x + 15 = -x^2$
       $x^2 - 8x + 15 = 0$ a = 1, b = -8, c = 15

1. $x^2 + 6x = -9$
     $x^2 + 6x + 9 = 0$ a = 1, b = 6, c = 9

2. $x^2 - 5 = 6x$
     $x^2 - 6x - 5 = 0$ a = 1, b = -6, c = -5

3. $8x + 12 = -x^2$
     $x^2 + 8x + 12 = 0$ a = 1, b = 8, c = 12

4. $m^2 - 4m = -4$
     $m^2 - 4m + 4 = 0$ a = 1, b = -4, c = 4

5. $y^2 - y = 20$
     $y^2 - y - 20 = 0$ a = 1, b = -1, c = -20

6. $a^2 - 30 = a$
     $a^2 - a - 30 = 0$ a = 1, b = -1, c = -30

7. $12x + 32 = x^2$
     $-x^2 + 12x + 32 = 0$ a = -1, b = 12, c = 32

8. $z^2 - 28 = -3z$
     $z^2 + 3z - 28 = 0$ a = 1, b = 3, c = -28

ANSWERS: Worksheet 31

Solve for x in the following quadratic equations.

1. $x^2 + 10x + 16 = 0$
    $(x + 8)(x + 2) = 0$
    $x = -8 \ or \ x = -2$

2. $x^2 + x - 12 = 0$
    $(x + 4)(x - 3) = 0$
    $x = -4 \ or \ x = 3$

3. $x^2 + 12x + 36 = 64$
    $(x + 6)(x + 6) = 64$
    $(x + 6)^2 = 64$
    $\sqrt{(x + 6)^2} = \pm\sqrt{64}$
    $x + 6 = \pm 8$
    $x = 2 \ or \ x = -14$

4. $a^2 + 15 = -8a$
    $a^2 + 8a + 15 = 0$
    $(a + 5)(a + 3) = 0$
    $a = -5 \ or \ a = -3$

5. $x^2 - 11x = -24$
    $x^2 - 11x + 24 = 0$
    $(x - 8)(x - 3) = 0$
    $x = 8 \ or \ x = 3$

6. $(x + 3)^2 = 49$
    $\sqrt{(x + 3)^2} = \pm\sqrt{49}$
    $x + 3 = \pm 7$
    $x = 4 \ or \ x = -10$

7. $y^2 = -3y - 2$
    $y^2 + 3y + 2 = 0$
    $(y + 1)(y + 2) = 0$
    $y = -1 \ or \ y = -2$

8. $25 = x^2 - 18x + 81$
    $x^2 - 18x + 81 = 25$
    $(x - 9)(x - 9) = 25$
    $(x - 9)^2 = 25$
    $\sqrt{(x - 9)^2} = \sqrt{25}$
    $x - 9 = \pm 5$
    $x = 4 \ or \ x = 14$

ANSWERS: Worksheet 32

Solve for x in the following quadratic equations.

1. $x^2 + 6x + 9 = 16$
   $(x+3)(x+3) = 16$
   $(x+3)^2 = 16$
   $\sqrt{(x+3)^2} = \pm\sqrt{16}$
   $x + 3 = \pm 4$
   $x = 1 \ \ or \ \ x = -7$

2. $x^2 + 8x + 16 = 121$
   $(x+4)(x+4) = 121$
   $(x+4)^2 = 121$
   $\sqrt{(x+4)^2} = \pm\sqrt{121}$
   $x + 4 = \pm 11$
   $x = 7 \ \ or \ \ x = -15$

3. $x^2 - 12x + 36 = 10$
   $(x-6)^2 = 10$
   $\sqrt{(x-6)^2} = \pm\sqrt{10}$
   $x - 6 = \pm\sqrt{10}$
   $x = 6 \pm \sqrt{10}$

4. $a^2 - 4a + 4 = 81$
   $(a-2)^2 = 81$
   $\sqrt{(a-2)^2} = \pm\sqrt{81}$
   $a - 2 = \pm 9$
   $a = -7 \ \ or \ \ a = 11$

5. $y^2 + 10y + 25 = 8$
   $(y+5)^2 = 8$
   $\sqrt{(y+5)^2} = \pm\sqrt{8}$
   $y + 5 = \pm\sqrt{8}$
   $y = -5 \pm \sqrt{8}$

6. $x^2 - 16x + 64 = 12$
   $(x-8)^2 = 12$
   $\sqrt{(x-8)^2} = \pm\sqrt{12}$
   $x - 8 = \pm\sqrt{12}$
   $x = 8 \pm \sqrt{12}$

7. $x^2 + 14x + 49 = 64$
   $(x+7)^2 = 64$
   $\sqrt{(x+7)^2} = \pm\sqrt{64}$
   $x + 7 = \pm 8$
   $x = 1 \ \ or \ -15$

8. $b^2 - 18b + 81 = 144$
   $(b-9)^2 = 144$
   $\sqrt{(b-9)^2} = \pm\sqrt{144}$
   $b - 9 = \pm 12$
   $b = 21 \ \ or \ \ b = -3$

ANSWERS: Worksheet 33

Solve the following quadratic equations, by completing the square with $\left(\frac{b}{2}\right)^2$.

1. $x^2 - 2x - 2 = 0$
   $x^2 - 2x = 2$
   $x^2 - 2x + 1 = 2 + 1$
   $(x - 1)^2 = 3$
   $\sqrt{(x-1)^2} = \pm\sqrt{3}$
   $x - 1 = \pm\sqrt{3}$
   $x = 1 \pm \sqrt{3}$

2. $x^2 - 2x - 10 = 0$
   $x^2 - 2x = 10$
   $x^2 - 2x + 1 = 10 + 1$
   $(x - 1)^2 = 11$
   $\sqrt{(x-1)^2} = \pm\sqrt{11}$
   $x - 1 = \pm\sqrt{11}$
   $x = 1 \pm \sqrt{11}$

3. $x^2 + 4x - 15 = 0$
   $x^2 + 4x = 15$
   $x^2 + 4x + 4 = 15 + 4$
   $(x + 2)^2 = 19$
   $\sqrt{(x+2)^2} = \pm\sqrt{19}$
   $x + 2 = \pm\sqrt{19}$
   $x = -2 \pm \sqrt{19}$

4. $\frac{1}{4}x^2 - \frac{2}{4}8x - \frac{6}{4}24 = \frac{1}{4}4$
   $x^2 - 2x - 6 = 1$
   $x^2 - 2x + 1 = 7 + 1$
   $x^2 - 2x + 1 = 8$
   $(x - 1)^2 = 8$
   $\sqrt{(x-1)^2} = \pm\sqrt{8}$
   $x - 1 = \pm\sqrt{8}$
   $x = 1 \pm \sqrt{8}$

ANSWERS: Worksheet 34

$$x = \frac{-b \pm \sqrt{b^2 - 4ac}}{2a}$$

1. $3x^2 - 2x - 5 = 0$

$$x = \frac{-(-2) \pm \sqrt{(-2)^2 - 4 \cdot 3 \cdot -5}}{2(3)}$$

$$x = \frac{2 \pm \sqrt{4 + 60}}{6}$$

$$x = \frac{2 \pm 8}{6}$$

$$x = \frac{10}{6} = \frac{5}{3} \quad or \quad x = \frac{-6}{6} = -1$$

2. $2x^2 - 5x + 3 = 0$

$$x = \frac{-(-5) \pm \sqrt{(-5)^2 - 4 \cdot 2 \cdot 3}}{2(2)}$$

$$x = \frac{5 \pm \sqrt{25 - 24}}{4}$$

$$x = \frac{5 \pm \sqrt{1}}{4}$$

$$x = \frac{5 \pm 1}{4}$$

$$x = \frac{5 + 1}{4} \qquad x = \frac{5 - 1}{4}$$

$$x = \frac{6}{4} \qquad x = \frac{4}{4}$$

$$x = \frac{3}{2} \quad or \quad x = 1$$

ANSWERS: Worksheet 34 page 2

3. $y^2 + 2y - 8 = 0$

$$x = \frac{-(2) \pm \sqrt{2^2 - 4(1)(-8)}}{2(1)}$$

$$x = \frac{-2 \pm \sqrt{4 + 32}}{2}$$

$$x = \frac{-2 \pm \sqrt{36}}{2}$$

$$x = \frac{-2 - 6}{2} \qquad\qquad x = \frac{-2 + 6}{2}$$

$$x = \frac{-8}{2} = -4 \qquad \textbf{or} \qquad x = \frac{4}{2} = 2$$

ANSWERS: Worksheet 35

1. $f(x) = x^2 + 2x - 35$
   $f(5) = 5^2 + 2(5) - 35$
   $f(5) = 25 + 10 - 35$
   $\boldsymbol{f(5) = 0}$

2. $f(x) = x^2 - 8x - 20$
   $f(-2) = (-2)^2 - 8(-2) - 20$
   $f(-2) = 4 + 16 - 20$
   $\boldsymbol{f(-2) = 0}$

3. $f(x) = x^2 - 3x - 30$
   $f(8) = 8^2 - 3(8) - 30$
   $f(8) = 64 - 24 - 30$
   $\boldsymbol{f(8) = 10}$

4. $f(x) = x^2 + 4x - 10$
   $f(3) = 3^2 + 4(3) - 10$
   $f(3) = 9 + 12 - 10$
   $\boldsymbol{f(3) = 11}$

5. $f(x) = x^2 + 4x - 22$
   $f(4) = 4^2 + 4(4) - 22$
   $f(4) = 16 + 16 - 22$
   $\boldsymbol{f(4) = 10}$

6. $f(x) = x^2 + \frac{1}{2}x - \sqrt{64}$
   $f(6) = 6^2 + \frac{1}{2}(6) - 8$
   $f(6) = 36 + 3 - 8$
   $\boldsymbol{f(6) = 31}$

7. $f(x) = x^2 + 9x - 84$
   $f(7) = 7^2 + 9(7) - 84$
   $f(7) = 49 + 63 - 84$
   $\boldsymbol{f(7) = 28}$

ANSWERS: Final Test

Solve the following quadratics by factoring. Check your answers by replacing x with your answer.

1. $x^2 - 2x - 24 = 0$
   $(x-6)(x+4) = 0$
   $\boldsymbol{x = 6 \ \ or \ \ x = -4}$

   Am I right? $x = 6$
   $6^2 - 2(6) - 24 = 0$
   $36 - 12 - 24 = 0$ **Yep!**
   Am I right? $x = -4$
   $4^2 - 2(-4) - 24 = 0$
   $16 + 8 - 24 = 0$ **Yep!**

2. $x^2 - 3x - 28 = 0$
   $(x-7)(x+4) = 0$
   $\boldsymbol{x = 7 \ \ or \ \ x = -4}$

   Am I right? $x = 7$
   $7^2 - 3(7) - 28 = 0$
   $49 - 21 - 28 = 0$ **Yep!**
   Am I right? $x = -4$
   $(-4)^2 - 3(-4) - 28 = 0$
   $16 + 12 - 28 = 0$ **Yep!**

3. $x^2 + 22x + 121 = 0$
   $(x+11)(x+11) = 0$
   $(x+11)^2 = 0$
   $x + 11 = \pm\sqrt{0}$
   $\boldsymbol{x = -11}$

   Am I right? $x = -11$
   $-11^2 + 22(-11) + 121 = 0$
   $121 - 242 + 121 = 0$ **Yep!**

4. $x^2 - x - 6 = 0$
   $(x-3)(x+2) = 0$
   $\boldsymbol{x = 3 \ \ or \ x = -2}$

   Am I right? $x = 3$
   $3^2 - 3 - 6 = 0$
   $9 - 9 = 0$ **Yep!**
   Am I right? $x = -2$
   $-2^2 + 2 - 6 = 0$
   $4 + 2 - 6 = 0$ **Yep!**

Continued on next page...

ANSWERS: Final Test page 2

Solve the following two quadratic equations, by completing the square. $\left(\frac{b}{2}\right)^2$
Check your work, by replacing x with your answers.

5. $x^2 - \frac{1}{2}x - 3 = 0$

   $x^2 - \frac{1}{2}x = 3$ $\left(\frac{b}{2}\right)^2 = \left(\frac{-\frac{1}{2}}{2}\right)^2 = -\frac{1}{2} \cdot \frac{1}{2} = -\frac{1}{4}$ and $\left(-\frac{1}{4}\right)^2 = \frac{1}{16}$

   $x^2 - \frac{1}{2}x + \frac{1}{16} = \frac{49}{16}$

   $\left(x - \frac{1}{4}\right)\left(x - \frac{1}{4}\right) = \frac{49}{16}$

   $\left(x - \frac{1}{4}\right)^2 = \sqrt{\frac{49}{16}}$

   $x - \frac{1}{4} = \pm\frac{7}{4}$

   $x = \frac{1}{4} \pm \frac{7}{4}$ $x = \frac{8}{4} = 2$ or $x = -\frac{6}{4} = -\frac{3}{2}$

6. $x^2 + 10x - 24 = 0$

   $x^2 + 10x = 24$ $\left(\frac{b}{2}\right)^2 = \left(\frac{10}{2}\right)^2 = 10 \div 2 = 5$ $(5)^2 = 25$

   $x^2 + 10x + 25 = 24 + 25$
   $(x + 5)(x + 5) = 49$
   $(x + 5)^2 = 49$
   $x + 5 = \pm 7$ $x = 2$ $x = -12$

Solve the following quadratics, by using the Quadratic Formula. $x = \frac{-b \pm \sqrt{b^2 - 4ac}}{2a}$

7. $2x^2 + 5x = 12$
   $2x^2 + 5x - 12 = 0$
   $x = \frac{-5 \pm \sqrt{(5)^2 - 4(2)(-12)}}{2(2)}$
   $x = \frac{-5 \pm \sqrt{25 + 96}}{4}$
   $x = \frac{-5 \pm \sqrt{121}}{4}$
   $x = \frac{-5 \pm 11}{4}$ $x = \frac{-16}{4} = -4$ or $x = \frac{6}{4} = \frac{3}{2}$

ANSWERS: Final Test page 3

8. $x^2 + 9x + 4 = 4$
    $x^2 + 9x + 4 - 4 = 4 - 4$
    $x^2 + 9x + 0 = 0$
    $x = \frac{-9 \pm \sqrt{9^2 - 4(1)(0)}}{2(1)}$
    $x = \frac{-9 \pm \sqrt{81 - 0}}{2}$
    $x = \frac{-9 \pm 9}{2}$ $\quad\quad x = \frac{0}{2} = 0 \quad$ or $\quad x = \frac{-18}{2} = -9$

9. $x^2 - 4x - 12 = 0$
    $(x - 6)(x + 2) = 0$
    $x = 6 \quad or \quad x = -2$

10. $f(x) = x^2 - 4x - 12$
    $f(1) = 1^2 - 4(1) - 12$
    $f(1) = 1 - 4 - 12$
    $f(1) = -15$

11. $f(x) = x^2 - 4x - 12$
    $f(2) = 2^2 - 4(2) - 12$
    $f(2) = 4 - 8 - 12$
    $f(2) = -16$

12. $f(x) = x^2 - 4x - 12$
    $f(-1) = -1^2 - 4(-1) - 12$
    $f(-1) = 1 + 4 - 12$
    $f(-1) = -7$

13. $f(x) = x^2 - 4x - 12$
    $f(4) = (4)^2 - 4(4) - 12$
    $f(4) = 16 - 16 - 12$
    $f(4) = -12$

ANSWERS: Final Test page 4

14. $f(x) = x^2 - 4x - 12$
     $f(5) = 5^2 - 4(5) - 12$
     $f(5) = 25 - 20 - 12$
     $\boldsymbol{f(5) = -7}$

15. $f(x) = x^2 - 4x - 12$
     $f(3) = 3^2 - 4(3) - 12$
     $f(3) = 9 - 12 - 12$
     $\boldsymbol{f(3) = -15}$

x	$f(x)$
6	0
-2	0
1	-15
2	-16
-1	-7
4	-12
5	-7
3	-15

ANSWERS: Final Test page 5

Fill in the numbers on the number lines and then plot the points you added to the table.

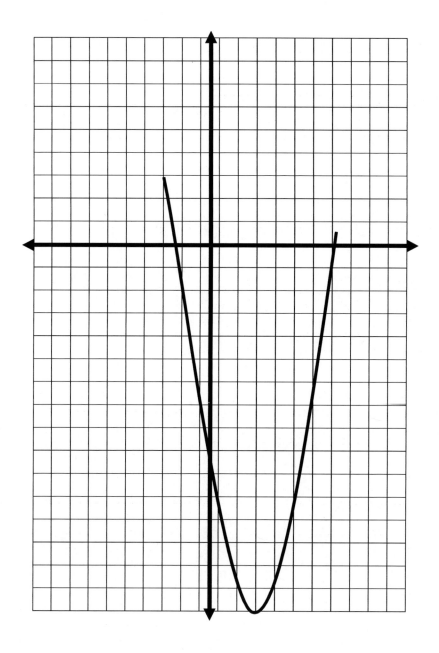

Made in the USA
Las Vegas, NV
01 April 2022